黑龙江嘉荫
晚白垩世植物群
Late Cretaceous flora from
Jiayin of Heilongjiang, China

Late Cretaceous flora from
Jiayin of Heilongjiang, China

黑龙江嘉荫
晚白垩世植物群

孙革　梁飞　杨涛　张淑芹　著

上海科技教育出版社

图书在版编目(CIP)数据

黑龙江嘉荫晚白垩世植物群/孙革等著. —上海：上海科技教育出版社,2020.12

ISBN 978-7-5428-7383-5

Ⅰ.①黑… Ⅱ.①孙… Ⅲ.①植物区系—嘉荫县—晚白垩世 Ⅳ.①Q948.523.54

中国版本图书馆CIP数据核字(2020)第194301号

责任编辑　伍慧玲
封面设计　汤世梁
版式设计　李梦雪

黑龙江嘉荫晚白垩世植物群

孙革　梁飞　杨涛　张淑芹　著

地图由中华地图学社授权使用,地图著作权归中华地图学社所有

出版发行	上海科技教育出版社有限公司	
	（上海市柳州路218号　邮政编码200235）	
网　　址	www.sste.com　www.ewen.co	
经　　销	各地新华书店	
印　　刷	上海华顿书刊印刷有限公司	
开　　本	787×1092　1/16	
印　　张	11.5	
插　　页	2	
版　　次	2020年12月第1版	
印　　次	2020年12月第1次印刷	
审 图 号	GS(2020)6877号	
书　　号	ISBN 978-7-5428-7383-5/N·1104	
定　　价	98.00元	

内容简介

　　黑龙江嘉荫地区晚白垩世中—晚期以恐龙为代表的生物群十分繁盛,含有丰富的植物化石。近十余年来,以我国古植物学家孙革为首的国际科研队在嘉荫晚白垩世中—晚期植物化石研究中取得了令人瞩目的重要成果,包括首次发现被子植物达勒比叶、水生被子植物嘉荫莲及首次发现葛赫叶的表皮构造等,新建立两个植物化石组合并确定了时代,确认了此间被子植物在植物群中的主导地位,并深入研究了被子植物与恐龙等协同演化关系等。

　　本书详细介绍了这些重大研究成果的发现过程,并在此基础上提出了许多极富价值的推论,特别是首次提出对我国东北晚白垩世植物群演替序列的新认识。上述成果对推动我国乃至东北亚地区晚白垩世植物群、相关地层及古地理环境等研究具有重要科学价值,也为相关地质古生物知识的科普做了有益的尝试。

目　录

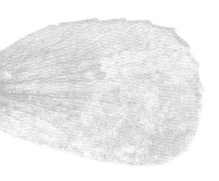

CONTENTS

前　言

　　嘉荫位于黑龙江省伊春地区东北部,毗邻中俄界河——美丽的黑龙江。这里依山傍水,景色迷人:墨绿色的群山绵延起伏,沿江伫立;淡蓝色的黑龙江波光潋滟,鱼钓舟摇;隔江向俄罗斯方向望去,沃野千里,一片金黄。这里辽阔、秀美、宁静,令人心旷神怡(图1)。

　　从恐龙等古生物化石的发现得知,约8000万~7000万年前,嘉荫是一处喧闹的"生物乐园",当时河湖密布、气候温暖,恐龙、龟、鱼等众多生物在此繁衍生息,湖中荷花等水生被子植物争艳,周边的山地被梧桐、水杉、银杏等高大树木组成的森林所覆盖。嘉荫神奇的古生物化石带我们走进距今约8000万~7000万年前、白垩纪中—晚期的远古世界。

　　嘉荫一直被誉为"中国龙乡",1902年,在嘉荫龙骨山发现了我国最早命名的恐龙化石——黑龙江满洲龙(*Mandschurosaurus amurensis*),这里的恐龙在我国又最晚灭绝,因此,嘉荫恐龙具有"一早、一晚"的特色。近年来,随着区域地质调查和古生物研究的深入,科学家在嘉荫及其邻区又发现了一大批新的恐龙化石,包括卡龙、乌拉嘎龙、黑龙、阿穆尔龙、大天鹅龙、昆都尔龙、克伯龙及暴龙类等10多个分类群,为研究我国及东北亚地区晚白垩世恐龙演化提供了一个理想的化石产地。此外,与恐龙伴生的大量植物化石等也相继被发现,特别如水生被子植物——嘉荫莲(荷花)等,为嘉荫这一化石宝地又增添了新的色彩。与此同时,随着古生态以及黑龙江省东、西部地层对比研究的开展,有关嘉

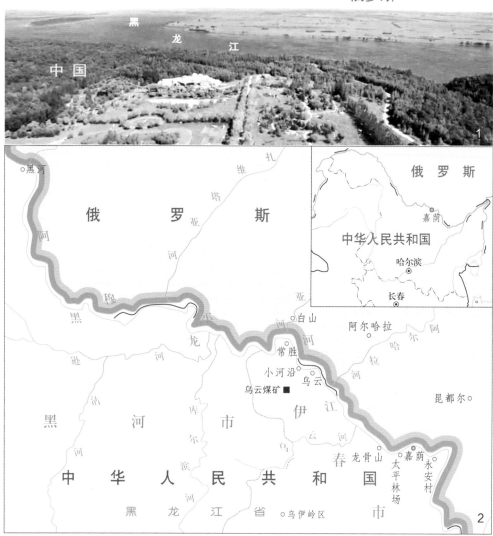

图1　嘉荫及其邻区地理位置

Fig. 1　Geographic map of Jiayin and its neighboring areas

1. 中国嘉荫恐龙地质公园与隔江的俄罗斯鸟瞰；
2. 嘉荫（红点示）及其邻区地理位置图；3. 嘉荫乌云小河沿黑龙江畔

1. Aerial view of National Dinosaur Geopark of Jiayin, China and Russian sides, separated by Heilongjiang (Amur) River; 2. Geographic map of Jiayin(the red dot) and its neighboring areas; 3. Beach of Heilongjiang (Amur) River in Xiaoheyan of Wuyun, Jiayin

荫晚白垩世地理、古气候及古生态等的研究被提上日程。特别是，近年来有关恐龙的深入研究表明，在嘉荫的恐龙化石中，约有95%属于植食的鸭嘴龙类，晚白垩世在嘉荫地区生长的繁茂的植物曾是恐龙生存与发展的物质来源。晚白垩世(距今约1.05亿~0.66亿年)是中生代最后一个时期，也是地球演化史中最动荡的时期之一。在这一漫长时间里，恐龙最终走向灭绝，全球植被也随之发生巨大的改变。那么，嘉荫晚白垩世植被究竟由哪些植物组成? 哪些植物曾上过恐龙的"餐桌"? 这些植物的兴衰是否曾影响了恐龙的发展与消亡? 恐龙与植物之间又是如何协同演化的? 这一系列科学问题已摆在科学家特别是古植物学家面前。

2002年起，由中国古植物学家孙革率领的国际科研队联合中、俄、德、美、英、比、日等国的科学家，共同在嘉荫地区开展了长达十余年的研究工作(图2)。除在恐龙及K-Pg地层界线研究中取得重要成果外，在嘉荫晚白垩世植物群研究方面也取得诸多新的进展，如晚白垩世重要被子植物达勒比叶(*Dalembia*)的发现、银杏类古气候定量研究，以及嘉荫莲、葛赫叶、卡波叶等一大批水生被子植物化石在嘉荫的新发现等。这些新成果不仅进一步揭示了嘉荫地区晚白垩世植物群的组成面貌，也为综合研究晚白垩世时期嘉荫及其邻区古地理、古气候、地层划分与对比，以及我国晚白垩世植物群演化发展等作出了贡献。及时总结上述成果、并将新成果介绍给业内同仁们，是作者撰写本书的主要初衷。此外，自2015年以来，本书作者承担了科技部"中国白垩系—古近系界线研究"及中国地质调查局"白垩系—古近系界线"这两项重要课题，课题工作也取得诸多新进展。应课题结题要求，须对嘉荫晚白垩世植物群的最新研究进行全面总结和汇报，这也是撰写本书的另一个原因。

我国最早对嘉荫晚白垩世植物群开展研究的是古植物学家张志诚(中国地调局沈阳地调中心，原沈阳地矿所)，他曾为嘉荫晚白垩世植物研究做了开创性工作。同时期研究者还有古植物学家郭双兴、陶君容等。近20年来由孙革(吉林大学/沈阳师范大学)领导的国际科研队专家对嘉荫晚白垩世植物群开展了深入研究，其中，以阿克米梯耶夫(M. Akhmetiev)院士、玛尔凯维奇(V. Markevich)及高洛夫涅娃(L. Golovneva)教授为首的俄罗斯专家，迪尔切(D. L. Dilcher)院士、约翰森(K. Johnson)及尼科斯(D. Nichols)教授等美国专家，德国孢粉学家阿什拉夫(A. R. Ashraf)，日本专家西田治文(Nishida H.)与铃木茂之(Suzuki S.)教授，以及我国专家孙春林、孙跃武、全成教授等，均为新一轮研究嘉

图2　吉林大学前校长刘中树接见国际科研队专家(2002年9月,长春)
Fig. 2　Ex-President and Prof. Liu Z. S. meeting experts of Int'l Research Team in Jilin University (Changchun, 2002)

前排右起:阿什拉夫(德),孙春林,迪尔切(美),陈竞捷(前黑龙江省国土厅副厅长),刘中树,阿克米梯耶夫院士(俄),凯金娜(俄),冈田博有(日),米家榕,柯珠尔(俄),王雅茹。2排左2:约翰森(美)。3排左起:孙革,董枝明

From right in front row: A. R. Ashraf, Sun C. L., D. L. Dilcher (NAS), Chen J. J. (ex-Vice-Director of DL-RH), Liu Z. S., M. Akhmetiev, T. Kezina, Okada H., Mi J. R., T. Kodrul, Wang Y. R. Left 2nd, 2nd row: K. Johnson. From left, 3rd row: Sun G., Dong Z. M.

荫晚白垩世植物群作出了重要贡献。因此,本书的研究成果是国内外专家集体智慧和汗水的结晶。

　　本书是我国首部系统、全面地报道黑龙江嘉荫地区晚白垩世植物群的科学专著,也是一部与嘉荫恐龙等研究有着密切联系的科普读物。全书共6章18节,首次全面报道了黑龙江嘉荫晚白垩世中—晚期植物群的组成、性质、时代及其古地理与古环境,除详细记述已发现的34属43种植物化石及进一步介绍相关的5个孢粉组合外,还讨论了嘉荫晚白垩世植物群与恐龙的协同演化及相关的嘉荫"K–Pg"界线,并介绍了作者对中国东北晚白垩世植物群发展序列研究的最新认识。此外,本书首次以科普形式介绍了嘉荫晚白垩世

"恐龙的餐桌"以及"嘉荫古生物学家雕塑园"等,书中还附有百余张彩色图片及英文全文。本书的问世不仅将丰富我国乃至国际白垩纪植物群研究的宝库,对研究我国及东北亚地区中生代地质/生物演化等也具有重要意义。

由于本书作者的水平和写作时间等局限,本书一定存在许多不尽理想的地方,特别是对以往有关嘉荫晚白垩世植物研究的资料尚未能全部收录或作进一步讨论等。作者期待未来有更新的有关嘉荫晚白垩世植物群的著作问世。

2020 年 6 月

嘉荫莲

Nelumbo jiayinensis

第一章

黑龙江嘉荫地区地质概况

1.1　地质概况

嘉荫位于黑龙江省伊春市北部(中心地理坐标130°32′44″E,48°53′29″N;图3, 1)。伊春地区连同黑龙江省同处于西伯利亚板块与华北板块之间的复合造山带,即兴蒙造山带,属古亚洲洋构造域东段与滨太平洋构造域交汇部位(李锦铁等,2009)。伊春地区最早的化石记录可追溯到距今5亿多年的寒武纪早期,在伊春市西林区的五星镇组曾发现寒武纪三叶虫(原叶尔伯虫,*Proerbia*)(段吉业、安素兰,2001),在伊春邻区的多宝山地区曾发现奥陶纪早、中期的海生生物化石(黑龙江省地质矿产局, 1993),说明寒武纪及奥陶纪中期,伊春地区曾是一片大海。奥陶纪中期(距今约4.7亿~4.6亿年)这里曾发生过加里东运动,有过大规模火山活动及岩浆活动,形成大量金属矿产。至泥盆纪早期,这里地壳隆升、地层缺失。至泥盆纪早中期(距今约4.1亿年)海水又涌进伊春地区,使这里呈现滨海–浅海环境。晚泥盆世—石炭纪本区再次隆起,海水退去。二叠纪中晚期(距今约2.7亿~2.52亿年)本区以陆相沉积为主,局部见海陆交互相沉积,在伊春红山地区晚二叠世地层中发现著名的安加拉型植物群。至晚二叠世末,伊春地区火山活动再次强烈出现。

　　进入中生代(距今2.52亿~0.66亿年),伊春地区已完全成为陆地。三叠纪和侏罗纪,太平洋板块与亚洲大陆板块碰撞频繁,造成这里主要分布花岗岩、各类火山岩,以及相关的火山沉积碎屑岩。进入白垩纪,伊春北部的嘉荫盆地已进入俄罗斯结雅–布列亚盆地的范畴(图3,2)。东起嘉荫乌拉嘎,西至嘉荫乌云,主要沉积了晚白垩世中、晚期(距今约8600万~6600万年)的地层,包括晚白垩世中期的永安村组(河流相夹湖

图3　伊春地区地质图及嘉荫盆地示意图

Fig. 3　Geological map of Yichun area and sketch map showing the Jiayin basin

1. 伊春地区地质图(据Sun,2011*);2. 嘉荫盆地示意图(据孙革等,2014);3. 界江远眺。

1. Geological map of Yichun area (Sun, 2011*); 2. Sketch map showing the Jiayin basin (after Sun et al., 2014); 3. Overlooking the Heilongjiang (Amur) River in Jiayin

* Sun G. 2011. General introduction to geology of Yichun area and the recent advance on the study of stratigraphy and paleontology in Jiayin. Key-lecture in Int'l Symposium on Geology & Paleontology in Yichun, China, Aug. 20, 2011. Yichun, China (Geological map quoted from Regional Geological Survey No.1, Heilongjiang Prov., China).

泊相),晚白垩世晚期的太平林场组(湖泊相)、渔亮子组(河流相)及富饶组(湖泊相)等。在永安村组及渔亮子组的杂色粗碎屑岩沉积中发现大量恐龙化石;在嘉荫乌云小河沿的钻孔中,发现富饶组与其上的古新世乌云组下部的白山头段呈整合接触。嘉荫盆地晚白垩世中期永安村组之下的沉积地层迄今尚未发现,但在嘉荫永安村以南的赵家店(130°33′05.5″E,48°44′49.1″N),孙革等前不久曾发现一套以酸性为主的火山岩,锆石U–Pb同位素测年为101.6±1.2 Ma,似说明在嘉荫地区晚白垩世早期及早中期的沉积可能缺失。在嘉荫,永安村组之下为早白垩世晚期的火山岩地层(宁远村组?),彼此呈不整合接触。

至古近纪(距今6600万年起),随着黑龙江省东部新的大型断陷不断形成和发展,包括嘉荫盆地在内的结雅–布列亚盆地不断扩大,古新世沉积了以乌云组为代表的河流–湖沼相含煤碎屑岩建造,形成大型褐煤田——乌云煤田。此后,随着新构造运动,黑龙江河谷形成,新近纪的沉积主要为河流相。进入第四纪,地壳运动以差异性升降为特点,此间总体山区抬升,平原区相对下沉。经历多次冰川活动后,大地逐渐夷平,气候逐渐转为温暖,已接近今天的面貌(孙革等,2014)。

1.2 晚白垩世地层

黑龙江嘉荫地区晚白垩世地层十分发育,主要包括(自下而上)永安村组、太平林场组、渔亮子组及富饶组(表1)。本文报道的植物大化石主要产于永安村组及太平林场组;恐龙化石主要产于渔亮子组,但在永安村组发现有恐龙足迹(董枝明等,2003);孢粉化石在上述4个组中均有产出。由于嘉荫盆地在宏观上隶属于以俄罗斯为主的结雅–布列亚盆地,晚白垩世地层在黑龙江左岸的俄罗斯境内的白山(Belay Gora)及阿尔哈拉(Akhara)等地也广泛发育(图1,1),这便为研究嘉荫与俄罗斯邻区的白垩纪地层对比以及白垩纪–古近纪地层(K-Pg)界线对比等创造了有利条件(孙革等,2014)。

1.2.1 永安村组(K_{2yn})

永安村组主要见于嘉荫县城以东、沿黑龙江右岸的永安村(图1)东山一带,主要由黄灰色砂岩、粉砂岩、泥岩夹砂砾岩透镜体及薄层石膏等组成,局部夹煤线,顶部见薄层凝灰岩,总厚度大于125 m,总体显示滨浅湖相夹河流相沉积。本组典型剖面位于嘉荫永安村东山(中心地理坐标:130°31′24″E,48°50′57″N),地层产状总体上为

表1　黑龙江嘉荫晚白垩世地层及与俄罗斯邻区对比(据孙革等,2014,有修改)

Table 1　Stratigraphy of Upper Cretaceous in Jiayin of Heilongjiang and correlation with its adjacent area in Russia (after Sun et al.,2014,with revision)

组 ╲ 地区　　年代			中国 黑龙江嘉荫		俄罗斯 结雅-布列亚盆地		
古新统	丹尼阶	上	乌云组	含煤段	查加扬群	上查加扬组 上	含煤亚组
古新统	丹尼阶	下	乌云组	白山头段	查加扬群	上查加扬组 上	砂岩亚组
上白垩统	马斯特里赫特阶	上	富饶组		查加扬群	中	中查加扬组(布列亚组)
上白垩统	马斯特里赫特阶	中	渔亮子组 (主要恐龙层)	上段	查加扬群	下	下查加扬组(乌杜楚坎组) (主要恐龙层)
上白垩统	马斯特里赫特阶	下	渔亮子组 (主要恐龙层)	下段	查加扬群	下	下查加扬组(乌杜楚坎组) (主要恐龙层)
上白垩统	坎潘阶		太平林场组		昆都尔组		上扎维金组
上白垩统	桑顿阶		永安村组		昆都尔组		上扎维金组
上白垩统	康尼亚克阶		?		博古昌组		上扎维金组
上白垩统	土伦阶		?		博古昌组		上扎维金组
上白垩统	赛诺曼阶				下扎维金组		

300°~330°∠10°~20°。本组顶部与上伏的太平林场组底部的油页岩层为整合接触,该油页岩层曾发现俞氏链叶肢介[*Halysestheria yui*(Chang)],显示与松辽盆地嫩江组底部相当(李罡等,2004);本组底部为以酸性火山岩为主的地层,黑龙江省地矿局认为,该火山岩为早白垩世宁远村组,并推测彼此为不整合接触(黑龙江省地质矿产局,1993)。前不久,孙革等在永安村以南的赵家店(130°33′05.5″E,48°44′49.1″N;属嘉荫县境内)发现一套以酸性为主的火山岩,锆石U-Pb同位素定年为101.6±1.2 Ma*,属早白垩世晚期的阿尔布期(Albian),可能属于黑龙江省地矿局认为的宁远村组。

永安村组产丰富的植物化石及少量动物化石。动物化石包括介形类、瓣鳃类及鸭嘴龙足迹化石——姜氏嘉荫龙足印(*Jiayinosauropus johnsoni* Dong et al.)等(图4,4)。植物化石包括丰富的植物大化石及孢粉化石。植物大化石经笔者研究主要为 *Parataxodium-Nelumbo*(准落羽杉-莲)组合,迄今已发现至少24属27种,包括蕨类 *Equisetum* sp., *Asplenium dicksonianum* Heer, *Arctopteris?* sp., *Cladophlebis* sp., *Glei-*

* 孙革.《中国白垩系—古近系界线研究》项目2017年度报告(内部资料).2017.

图4　嘉荫永安村东山永安村组

Fig. 4　Yong'ancun Formation in East Hill of Yong'an village, Jiayin

1、2. 永安村组地层剖面;3. 砂砾岩透镜体;4. 姜氏嘉荫龙足印

1, 2. Geological sections of Yong'ancun Formation; 3. Sand-conglomerate lens; 4. Dinosaur track *Jiayino-sauropus johnsoni*

chenites sp.;银杏类*Ginkgo adiantoides*（Ung.）Heer, *G. pilifera* Samylina;松柏类*Parataxodium* sp., *Metasequoia disticha*（Heer）Miki, *Sequoia* sp., *Cupressinocladus sveshnikovae* Ablajev, *Elatocladus* sp. 2;被子植物*Dalembia jiayinensis* Sun et Golovneva, *Trochodendroides arctica*（Heer）Berry, *Nyssidium arcticum*（Heer）Iljinskaja, *Platanus* sp., *Cobbania corrugata*（Lesq.）Stockey et al., *Nelumbo jiayinensis* Liang et al., *Quereuxia angulata*（Newb.）Krysht., 等（孙革等, 2014;梁飞、孙革, 2015;Sun et al., 2016;Liang et al., 2018）。孢粉化石为*Kuprianipollis santaloides-Duplosporis borealis*组合,时代为晚白垩世桑顿期（Santonian）（Markevich et al., 2011）。

1.2.2　太平林场组（K_{2tp}）

太平林场组主要出露于黑龙江右岸的嘉荫太平林场（图1）一带（中心地理坐标:130°14′58″E, 48°51′31″N）、嘉荫—乌云公路南侧以及嘉荫县城东南的"四号靶场"等地。太平林场组主要由黄灰、绿灰色砂岩、粉砂岩泥岩及油页岩组成,地层走向近NNE,倾角为8°~12°;顶部的粉砂岩与渔亮子组的砂砾岩为平行不整合接触;本组底部的油页岩层与永安村组为整合接触;全组总厚635 m以上,总体上显示湖泊相沉积(图5)。据李罡等(2004)研究,在嘉荫永安村以东的黑龙江岸边,在永安村组之上、太平林场底部的黑灰色页岩中发现俞氏链叶肢介 *Halysestheria yui* (Chang)(=青岗链叶肢介 *Halysestheria qinggangensis* Zhang et Chen),该种为嫩江组的标准化石,因此认为太平林场组在区域上可与松辽盆地晚白垩世嫩江组对比。此外,本书作者前不久在嘉荫四号靶场的太平林场组新发现长头松花江鱼(*Sungarichthys longicephalus* Takai)(图5,5、6);该化石以往曾发现于吉林前郭伏龙泉组(=嫩江组),这一新发现为嘉荫太平林场组与松辽盆地嫩江组的对比也提供了重要的动物化石证据。

图5 嘉荫晚白垩世太平林场组

Fig. 5 Upper Cretaceous Taipinglinchang Formation in Jiayin

1~3. 太平林场组剖面;4~6. 四号靶场剖面及其长头松花江鱼化石

1–3. Sections of Taipinglinchang Formation; 4–6. Section in Target Range No. 4 and its fish fossils of *Sungarichthys longicephalus* Takai

太平林场组产出的植物化石十分丰富,太平林场组植物组合被孙革等命名为 *Metasequoia-Trochodendroides-Quereuxia*(水杉-似昆栏树-葛赫叶)组合(Sun et al., 2007, 2011;孙革等,2014)。该组合迄今已发现植物化石30属38种,包括:苔藓类

Thallites sp.；蕨类 *Equisetum* sp.，*Asplenium dicksonianum* Heer，*Cladophlebis* sp.；银杏类 *Ginkgo adiantoides*（Ung.）Heer，*G. pilifera* Samylina；松柏类 *Parataxodium* sp.，*Metasequoia disticha*（Heer）Miki，*Sequoia* sp.，*Larix* sp.，*Gryptostrobus* sp.，*Pityophyllum* sp.，*Pityospermum* sp.，*Elatocladus* spp.1，2；被子植物 *Araliaephyllum?* sp.，*Arthollia tschernyschewii* Golovneva，Sun et Bugdaeva，*A. orientalia*（Zhang）Gol.，*Celastrinites kundurensis*（Konstanov）Gol.，Sun et Bugd.，*Platanus multinervis* Zhang，*P. sinensis* Zhang，*Platanus* sp.，*Trochodendroides arctica*（Heer）Berry，*T. lanceolata* Golovneva，*T. smilacifolia* Zhang，*T. taipinglinchanica* Golovneva，Sun et Bugdaeva，*Viburnophyllum* sp.，*Quereuxia angulata*（Newb.）Krysht.，*Cobbania corrugata*（Lesq.）Stockey et al.等（Sun et al.，2007，2011；孙革等，2014）。孢粉化石为 *Aquilapollenites conatus*-*Podocarpidites multesimus* 组合，时代显示晚白垩世坎潘期（Campanian）（Markevich et al.，2006，2011；孙革等，2014）。

1.2.3　渔亮子组（K_{2yl}）

　　渔亮子组主要分布于黑龙江右岸的嘉荫龙骨山（渔亮子村以东，见图1，中心地理位置坐标：130°13′40″E，48°51′30″N）及乌拉嘎等地。渔亮子组分为上、下两段：下段主要出露于嘉荫龙骨山，由灰绿、灰黄色砾岩、含砾砂岩、砂岩和泥岩等组成，厚约287 m，产满洲龙、卡龙等大量恐龙化石，孢粉化石显示为 *Aquilapollenites amygdaloides*-*Gnetaceaepollenites evidens* 组合（组合Ⅲ），时代为马斯特里赫特早期；上段主要出露于嘉荫乌拉嘎，由一套灰紫色砾岩及砂岩等组成，产董氏乌拉嘎龙（*Wulagasaurus dongi* God. et al.）及鄂伦春黑龙（*Salaliyania elunchunorum* God. et al.）等大量恐龙化石（Godefroit et al.，2011）（图6），孢粉化石显示 *Wodehouseia aspera*-*Parviprojectus amurensis* 组合（组合Ⅳ），时代为马斯特里赫特中期（Markevich et al.，2009；孙革等，2014）。渔亮子组普遍发育有斜层理和交错层理，沉积类型主要为河流相，该组底部平行不整合于太平林场组之上，该组与上覆富饶组的接触关系不明。

　　渔亮子组含丰富的以满洲龙为代表的恐龙等化石，这为该组时代确定提供了有力证据。该组发现的霸王龙类的威胁阿尔伯塔龙（*Albertosaurus periculosus* Riabinin）在北美产于加拿大阿尔伯塔上白垩统，时代为马斯特里赫特期（Russell，1970）。根据嘉荫地区 *Mandschurosaurus*-*Albertosaurus* 恐龙动物组合及孢粉组合，渔亮子组的时代目前多数认为属于马斯特里赫特早—中期（Markevich et al.，2006，2011；Sun et al.，2011；孙革等，2014），但也有恐龙专家认为其属于马斯特里赫特中晚期或晚期。

图6　嘉荫晚白垩世渔亮子组

Fig. 6　Upper Cretaceous Yuliangzi Formation of Jiayin

1. 龙骨山；2. 满洲龙；3. 渔亮子组下段剖面；4、5. 乌拉嘎渔亮子组上段及恐龙化石埋藏；6~8. 渔亮子组恐龙，6为阿穆尔龙（复原图），7为大天鹅龙（右）与阿穆尔龙，8为乌拉嘎龙

1. Longgushan; 2. Dinosaur *Mandschurosaurus*; 3. Lower Member of Yuliangzi Formation in Longgushan; 4, 5. Upper Member of Yuliangzi Formation in Wulaga and its dinosaur taphonomy; 6–8. Dinosaurs in Yuliangzi Formation: 6. *Amursaurus* (reconstruction); 7. *Olorotitan* (right) and *Amursaurus*; 8. *Wulagasaurus*

1.2.4　富饶组(K_{2fr})

　　富饶组是嘉荫晚白垩世年代最晚的一个组，也是研究嘉荫K-Pg界线最关键的地层之一。富饶组主要由深灰色泥岩及细、粉砂岩组成，局部见凝灰质粉砂岩等；主要见于嘉荫乌云小河沿钻孔，厚度大于136 m，代表一套湖相沉积（孙革等，2014）。富饶组是1981年由黑龙江省第一区调队所建，该组原含义是一套河湖三角洲相细碎屑沉积，主要由深灰色粉砂岩、砂岩、凝灰质角砾岩及酸性凝灰岩等组成，主要见于黑龙江东岸原富饶公社小河沿村低缓丘陵的陡坎一带，该组与下伏产恐龙化石的渔亮子组接触关系不明（黑龙江省地质矿产局，1993）；孢粉学家刘牧灵认为该组时代"可能为白垩纪—第三纪过渡"（刘牧灵，1990）。自2002年起，孙革等对富饶组重新进行了研究，根据新发现的古生物化石、同位素测年等证据，将出露于乌云小河沿至白山头一

带原"富饶组上部"地层(包括酸性凝灰岩、黑褐色细粉砂岩、炭质泥岩夹薄煤)从原含义的"富饶组"划分出来,重新命名为"白山头段",隶属乌云组下段,时代属古新世;原"富饶组"的中下部(即整合于白山头段之下的一套粉砂岩、泥岩及砂岩等)为修订含义后的富饶组;原"富饶组"从此解体(Sun et al.,2002;孙革等,2003,2005)。

　　然而,由于新划分的富饶组未能在地表完整出露,结合找寻K-Pg界线的需要,孙革等在2005~2008年先后于嘉荫乌云小河沿村以南实施了三个钻孔(XHY-2005,XHY-2006,XHY-2008),揭示了富饶组的组成(图7)。据钻孔岩心所含丰富的孢粉化石,结合古地磁、地球化学及同位素测年等综合研究,确认以XHY-2006钻孔岩心22.00~22.05 m 的地层为富饶组之顶及K-Pg界线,富饶组的时代为晚白垩世最末期(马斯特里赫特晚期),富饶组之上与古新统乌云组下段(白山头段)底部为整合接触。上述三个钻孔标定的富饶组上界即为白垩纪—古近纪地层(K-Pg)界线(Sun et al.,2011;孙革等,2014)(图7)。

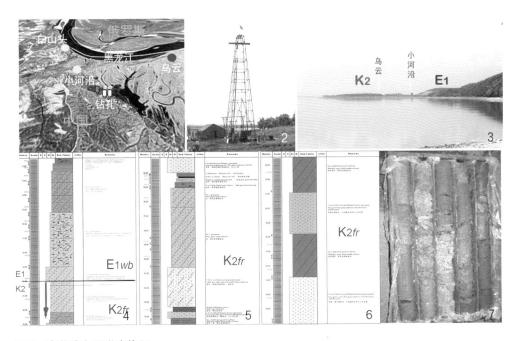

图7　嘉荫晚白垩世富饶组

Fig. 7　Upper Cretaceous Furao Formation in Jiayin

1~3.乌云小河沿及小河沿K-Pg钻孔位置示意;4~7.富饶组XHY-2005钻孔柱状图及岩心(据孙革等,2014)

1–3. Geographic positions of Xiaoheyan of Wuyun and the boreholes of K-Pg in Xiaoheyan; 4–7. Stratigraphic column of Furao Formation, shown by red point below in borehole XHY-2005, and its core rocks (after Sun et al., 2014)

迄今尚未在富饶组发现保存完好的植物大化石，但富饶组的孢粉化石十分丰富，以 *Aquilapollenites conatus-Pseudoaquilapollenites striatus* 组合（组合Ⅴ）为代表。经玛尔凯维奇等研究，富饶组已发现至少29属49种孢粉化石，其中，孢子以单缝孢 *Laevigatosporites* 为主，花粉中 *Aquilapollenites*（鹰粉）多样性强，至少10种以上。孢粉化石重要分类群主要包括 *Marsypiletes cretacea*，*Tricolpites variexinus*，*Aquilapollenites stelkii*，*A. proceros*，*A. striatus*，*A. rigidus*，*Intergricorpus bellum*，*Pseudointegricorpus clarireticulatus* 等。其中，时代仅限于马斯特里赫特晚期的包括 *Aquilapollenites stelkii*，*A. conatus*，*Pseudointegricorpus clarireticulatus*，*Marsypiletes cretacea*，*Integricorpus bellum* 等。上述孢粉化石较精确地指示，富饶组的时代为马斯特里赫特晚期（late Maastrichtian）（Markevich et al., 2006, 2009, 2011; 孙革等, 2014）。

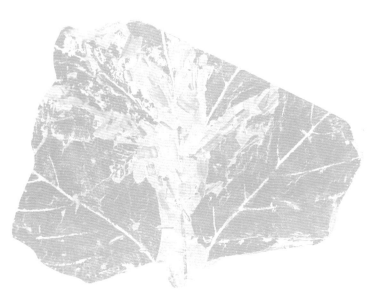

第二章
嘉荫晚白垩世植物群

2.1　植物群的组成

晚白垩世(距今1.05亿~0.66亿年),全球气候总体上偏高温和干旱,但在滨太平洋的俄罗斯远东及我国东北黑龙江东部一带,由于一些低缓的山间盆地距海不远,受海洋性温暖湿润气候等影响,植被总体上仍然繁茂(Herman,2002,2011;孙革等,2014)。从在嘉荫已发现的大量植物化石看,晚白垩世中—晚期(桑顿期—坎潘期,距今约8600万~7200万),这里曾生长着繁茂的、类似于现今我国长江流域一带低矮山地暖温带的森林。嘉荫晚白垩世中—晚期植物群以苔藓类、蕨类、银杏类、松柏类及被子植物为主,其中,被子植物已在植物群中占主导地位(Sun et al.,2007;孙革等,2014)。

相对于早白垩世植物群而言,黑龙江嘉荫晚白垩世植物群中的蕨类及银杏类等已大大减少,茨康类已近绝迹;在嘉荫迄今尚未发现苏铁类化石。总体上,在嘉荫晚白垩世中—晚期植物群中,松柏类仍占较大比例(约占27%),出现水杉(*Metasequoia*)、红杉(*Sequoia*)、水松(*Glyptostrobus*)等新类群;被子植物已得到较大发展且分异度大大增高,出现了似昆栏树(*Trochodendroides*)、似南蛇藤(*Celastrinites*)、达勒比叶(*Dalembia*)及悬铃木(*Platanus*)等高大阔叶落叶乔木被子植物,以及大量水生被子植物,如葛赫叶(*Quereuxia*)、卡波叶(*Cobbania*)及莲(*Nelumbo*)等,许多被子植物类群已接近其相关的现生的类群(张志诚,1984;全成,2006;Sun et al.,2007,2011;Golovneva et al.,2008;Quan & Sun,2008;孙革等,2014;梁飞、孙革,2015;Sun et al.,2016;Liang et al.,2018)。需要指出的是,自晚白垩世早期的赛诺曼期(Cenomanian)开始,我国东北黑龙江东部地区的晚白垩世植物群中,已出现高大的阔叶落叶被子植物悬铃木类(如悬铃木 *Platanus*)等植物,化石首见于黑龙江牡丹江-七台河地区的猴石沟组上部(张志诚,1981;Sun et al.,2019),并在黑龙江西部的松辽盆地东缘(如安达等地)的泉头组及姚家组等广泛分布(郑少林、张莹,1994);在嘉荫晚白垩世中—晚期植物群中,悬铃木类植物继续繁盛。

嘉荫晚白垩世中—晚期植物群迄今已发现34属43种植物大化石,包括苔藓类1属1种,有节类1属1种,真蕨类4属4种,银杏类1属3种,松柏类11属12种,被子植物15属21种,分类不明的种子化石1属1种。各分类群(种级)的组成比例为:苔藓类约占2.2%,蕨类约占11.6%,银杏类约占7.0%,松柏类约占27.0%,被子植物约占50.0%,分类不明的种子化石约占2.2%。

2.1.1 植物大化石

嘉荫晚白垩世植物群的组成大体上可划分为晚白垩世中期和晚期两个组合(Sun et al.,2007;孙革等,2014),包括:①永安村组植物组合(准落羽杉-莲组合,*Parataxodium-Nelumbo* Assemblage;时代为桑顿期,Santonian);②太平林场组植物组合(水杉-似昆栏树-葛赫叶组合,*Metasequoia-Trochodendroides-Quereuxia* Assemblage;时代为坎潘期,Campanian)。植物群的具体组成及其地质分布参见下表(表2)。

① 永安村组植物组合(准落羽杉-莲组合,*Parataxodium-Nelumbo* Ass.)

主要产于嘉荫永安村东山沿黑龙江右岸出露的永安村组,迄今已发现24属27种以上,主要包括:蕨类 *Equisetum* sp.,*Asplenium dicksonianum* Heer,*Arctopteris* sp.,*Cladophlebis* sp.,*Gleichenites* sp.;银杏类 *Ginkgo adiantoides*(Ung.)Heer,*G. pilifera* Samylina;松柏类 *Cupressinocladus sveshnikovae* Ablaev,*Metasequoia disticha*(Heer)Miki,*Sequoia* sp.,*Parataxodium* sp.,*Elatocladus* sp. 2;被子植物 *Dalembia jiayinensis* Sun et Golovneva,*Menispermites* sp.,*Trochodendroides arctica*(Heer)Berry,*Nyssidium arcticum*(Heer)Iljinskaja,*Platanus* sp.,*Viburnophyllum* sp.,*Dicotylophyllum* sp.,*Quereuxia angulata*(Newb.)Krysht.,*Cobbania corrugata*(Lesq.)Stockey et al.,*Nelumbo jiayinensis* Liang et al. 等(孙革等,2014;梁飞、孙革,2015;Sun et al.,2016;Liang et al.,2018)。根据孢粉

表2　嘉荫晚白垩世植物群组成及地质分布

Table 2　Composition of Late Cretaceous flora in Jiayin and their geological horizons

序号	分类群 Taxa	永安村组 (K_{2yn}) 标本数	太平林场组 (K_{2tp}) 标本数	标本总数
	Bryophytes 苔藓类			
1	*Tallites* sp. 似叶状体（未定种）	0	4	4
	Equisetales 有节类			
2	*Equisetum* sp. 木贼（未定种）	4	8	12
	Filicales 真蕨类			
3	*Asplenium dicksonianum* 迪克逊铁角蕨	69	20	89
4	*Arctopteris* sp. 北极蕨（未定种）	0	2	2
5	*Cladophlebis* sp. 枝脉蕨（未定种）	10	2	12
6	*Gleichenites* sp. 似里白（未定种）	10	0	10
	Ginkgoales 银杏类			
7	*Ginkgo adiantoides* 似铁线蕨型银杏	20	10	30
8	*Ginkgo pilifera* 具毛银杏	3	5	8
9	*Ginkgo* sp. 银杏（未定种）	15	5	20
	Coniferales 松柏类			
10	*Parataxodium* sp. 准落羽杉（未定种）	20	5	25
11	*Taxodium olrikii* 奥尔瑞克落羽杉	0	4	4
12	*Metasequoia disticha* 二列水杉	30	50	80
13	*Sequoia* sp. 红杉（未定种）	6	4	10
14	*Gryptostrobus* sp. 水松（未定种）	23	2	25
15	*Larix* sp. 落叶松（未定种）	0	1	1
16	*Pityospermum minutum* 小松型籽	4	2	6
17	*Pityophyllum* sp. 松型叶（未定种）	2	2	4
18	*Cupressinocladus sveshnikovae* 司氏柏型枝	5	10	15
19	*Elatocladus* sp. 1 枞型枝（未定种1）	0	1	1
20	*Elatocladus* sp. 2 枞型枝（未定种2）	1	6	7
21	cf. *Podocarpus tsagajanicus* 查加扬罗汉松（比较属种）	0	1	1
	Angiospermae 被子植物			
22	*Araliaephyllum?* sp. 惚木叶?（未定种）	2	0	2
23	*Arthollia tschernyschewii* 车氏阿朔叶	0	5	5
24	*Arthollia orientalis* 东方阿朔叶	0	4	4
25	*Celastrinites kundurensis* 昆都尔似南蛇藤	0	7	7

（续表）

序号	分类群 Taxa	永安村组 （K_{2yn}） 标本数	太平林场组 （K_{2tp}） 标本数	标本总数
26	*Dalembia jiayinensis* 嘉荫达勒比叶	20	1	21
27	*Menispermites* sp. 拟蝙蝠叶（未定种）	2	0	2
28	*Nordenskioidea* cf. *Borealis* 北方诺登斯基果（比较种）	0	1	1
29	*Nyssidium arcticum* 北极准紫树果	10	3	13
30	*Platanus densinervis* 密脉悬铃木	0	1	1
31	*Platanus sinensis* 中华悬铃木	0	8	8
32	*Platanus* sp. 悬铃木（未定种）	4	2	6
33	*Trochodendroides arctica* 北极似昆栏树	76	10	86
34	*Trochodendroides lanceolata* 披针形似昆栏树	0	20	20
35	*Trochodendroides microdentatus* 微尖似昆栏树	0	2	2
36	*Trochodendroides taipinglinchanica* 太平林场似昆栏树	0	3	3
37	*Viburnum* cf. *contortum* 扭曲荚蒾（比较种）	0	1	1
38	*Viburnophyllum* sp. 荚蒾叶（未定种）	0	1	1
39	*Cabbania corrugata* 褶皱卡波叶	39	10	49
40	*Nelumbo jiayinensis* 嘉荫莲	20	7	27
41	*Quereuxia angulata* 具棱葛赫叶	30	100	130
42	*Dicotylophyllum* sp. 双子叶植物叶（未定种）	2	0	2
Seeds indet. systematically 分类不明的种子				
43	*Carpolithus* sp. 石籽（未定种）	5	2	7
	化石标本数合计	432	332	764

化石意见，时代为晚白垩世桑顿期（Markevich et al.，2011；孙革等，2014）。

② 太平林场组植物组合（水杉-似昆栏树-葛赫叶组合，*Metasequoia-Trochodendroides-Quereuxia* Ass.）

主要产于嘉荫太平林场一带沿黑龙江右岸出露的太平林场组，迄今已发现植物化石30属38种，包括：苔藓类 *Thallites* sp.；蕨类 *Equisetum* sp.，*Asplenium dicksonianum* Heer，*Cladophlebis* sp.；银杏类 *Ginkgo adiantoides*（Ung.）Heer，*G. pilifera* Samylina；松柏类 *Taxodium olrikii*（Heer）Brown，*Metasequoia disticha*（Heer）Miki，*Sequoia* sp.，*Pityophyllum* sp.，*Pityospermum* sp.，*Glyptostrobus* sp.，*Larix* sp.，*Elatocladus* spp. 1，2；被子植物 *Araliaephyllum?* sp.，*Arthollia orientalis*（Zhang）Golovneva，*A. tschernyschewii*（Kostanov）Golovneva，Sun et Bugdaeva，*Celastrinites kundurensis* Gol.，Sun et Bugd.，*Platanus densi-*

nervis Zhang，*P. sinensis* Zhang，*Platanus* sp.，*Trochodendroides arctica*（Heer）Berry，*T. taipinglinchanica* Gol.，Sun et Bugd.，*T. lanceolata* Gol.，*T. microdentatus*（Newb.）Krysht.，*Viburnum* cf. *contortum* Berry，*Viburnophyllum* sp.，*Quereuxia angulata*（Newb.）Krysht.，*Cobbania corrugata*（Lesq.）Stockey et al. 等（Sun et al.，2007，2011；Golovneva et al.，2008；孙革等，2014）。据孢粉化石看，时代为晚白垩世坎潘期（Markevich et al.，2011；孙革等，2014）。

需要说明的是，为植物系统分类描述的方便，本书将上述永安村组及太平林场组这两个组合的植物化石合并在一起，按分类系统加以描述，产出层位分别加以注明。此外，对前人报道的个别分类群（如张志诚，1984），本书未再详细描述，仅在该分类群的后面加"＊"号予以注明，其原始描述可查阅原作者论文。

以下为化石简要描述。

苔藓类 Bryophytes

似叶状体（未定种）*Thallites* sp.（图8，1）

苔型叶，总体呈楔形；叶小，两歧分支2~3次；裂片长线型，每枚裂片长4~5 mm，宽1~1.5 mm，表面平，具单一中脉，顶端钝圆。生殖部分未保存。

张志诚（1984）描述的 *Thallites jiayinensis* 实际上是葛赫叶（*Quereuxia*）的沉水叶。对此，本文作者及全成曾予以纠正（全成，2006；孙革等，2014）。

产出层位：太平林场组（K$_{2tp}$）。

蕨类 Filiales

有节类 Equisetales

木贼（未定种）*Equisetum* sp.（图8，2、3）

化石保存为不完整的茎及叶鞘部分。茎较细，宽0.8~1.0 cm，节部具叶鞘、略膨凸，叶鞘长1.2~1.5 cm，具叶约10枚。每枚叶宽1~2 mm，游离叶部分长4~6 mm，顶部渐尖。生殖部分未保存。

产出层位：永安村组（K$_{2yn}$）及太平林场组（K$_{2tp}$）。

真蕨类 Filicales

迪克逊铁角蕨 *Asplenium dicksonianum* Heer（图8，4~7）

化石多保存不完整的羽片部分。末二级羽片近披针形至菱形，长大于3~4 cm，宽

1.5~2 cm,羽轴细,宽约 1 mm。末级羽片近披针形至菱形,长 1.5~2 cm,宽 4~5 mm,顶端略尖,多以 30°~40° 自末二级羽轴上伸出;小羽片形态与羽片形态近似。生殖部分未保存。

产出层位:永安村组(K$_{2yn}$)及太平林场组(K$_{2tp}$)。

枝脉蕨(未定种)*Cladophlebis* sp. (图 8、8、9)

化石多保存为不完整羽片,羽片近长线形或长披针形,长大于 8 cm,宽约 2 cm,羽轴细,宽约 1 mm。小羽片近长方形至宽椭圆形,亚互生,长 0.8~1 cm,宽 4~5 mm,顶端钝尖,中脉较直,侧脉分叉 1~2 次。生殖部分未保存。

产出层位:永安村组(K$_{2yn}$)及太平林场组(K$_{2tp}$)。

似里白(未定种)*Gleichenites* sp. (图 8、10、11)

化石保存为不完整蕨叶,叶形体小,叶轴相对略粗;羽片近长线形至伸长披针形,长 0.8~1.2 cm,宽约 0.3 cm,羽轴相对略粗,宽约 0.5 mm。小羽片近宽椭圆形,长 2~3 mm,宽 1~1.5 mm,顶端钝圆;近对生,密集排列于羽轴两侧,叶脉似羽状,不很清

图 8　嘉荫晚白垩世苔藓类及蕨类

Fig. 8　Late Cretaceous mosses and ferns of Jiayin

1. 似叶状体(未定种),RCPS-TP2-26;2、3. 木贼(未定种),Yn-02-001,Yn-02-002;4~7. 迪克逊铁角蕨,Yn-B-1e,Yn-B-543,Yn-B-544,Yn-02-006;8、9. 枝脉蕨(未定种),Yn-B-530,YX-202;10、11. 似里白(未定种),Yn-B-566,Yn-B-560

1. *Thallites* sp., RCPS-TP2-26; 2, 3. *Equisetum* sp., Yn-02-001, Yn-02-002; 4~7. *Asplenium dicksonianum*, Yn-B-1e, Yn-B-543, Yn-B-544, Yn-02-006; 8, 9. *Cladophlebis* sp., Yn-B-530, YX-202; 10, 11. *Glechenites* sp., Yn-B-566, Yn-B-560

晰。生殖部分未保存。

产出层位:永安村组(K_{2yn})。

裸子植物 Gymnospermae

银杏类 Ginkgoales

银杏类是嘉荫晚白垩世植物群的重要组成部分。在嘉荫已发现的12属15种裸子植物中,含银杏类1属3种,包括 *Ginkgo adiantoides* (Ung.) Heer(似铁线蕨型银杏),*G. pilifera* Samylina(具毛银杏)以及 *Ginkgo* sp. 银杏(未定种)等。银杏类属种不多,但化石数量较大,在永安村组及太平林场组均有分布,而且大多保存完好的角质层。这些化石角质层为研究嘉荫地区银杏类化石分类、晚白垩世古气候及古地理环境等发挥了重要作用(全成,2006;公繁浩,2007;Sun et al.,2007;Quan & Sun,2008;Quan et al.,2009;孙革等,2014)。嘉荫晚白垩世银杏类虽与现生银杏(*Ginkgo biloba*)在形态结构等方面已较为接近,但仍有其特征(图9)。

似铁线蕨型银杏 *Ginkgo adiantoides* (Unger) Heer (图9,1~8;图10)

叶半圆至扇形,长3~4 cm,宽5~6 cm;叶缘全缘或略波状,基部宽楔形,叶脉自基部伸出,呈二歧分叉。叶角质层下气孔式(hypostomatal)。上表皮脉路区普通表皮细胞通常3~5列,多长方形或梭形,大小为71~122 μm ×12~18 μm;垂周壁直或波状弯曲;平周壁每个细胞中央具一乳突。脉间区普通表皮细胞呈不规则四边形或多边形,大小为32~50 μm ×18~35 μm;垂周壁波状弯曲;平周壁较平,偶见乳突。

下表皮脉路区由4~6列伸长细胞组成;垂周壁直或微弯;平周壁具乳突,呈串珠状,偶见毛。脉间区普通表皮细胞形态不明显,每个细胞中央具一乳突。气孔器呈单环式,大小约17~26 μm×10~13 μm;孔缝无定向;保卫细胞下陷,内缘加厚;副卫细胞4~7个,乳突状角质加厚强烈,常掩盖孔缝(图10,2~8)。

当前材料与Samylina(1963)描述的本种西伯利亚标本(Samylina,1963,第96页,图版22,图8、图9)在外形特征上十分相似,与她报道的本种表皮构造特征(Samylina,1967)也基本一致。

产出层位:永安村组(K_{2yn})及太平林场组(K_{2tp})。

具毛银杏 *Ginkgo pilifera* Samylina (图9,9、10;图11)

化石多保存较为完整的叶片,叶半圆形至扇形,长约3 cm,宽约6 cm;全缘或略波状,偶见中央微缺;基部为宽楔形,具一长柄,柄长大于2.5 cm,宽约2 mm;叶脉自基部伸出,呈二歧分叉。角质层多保存。

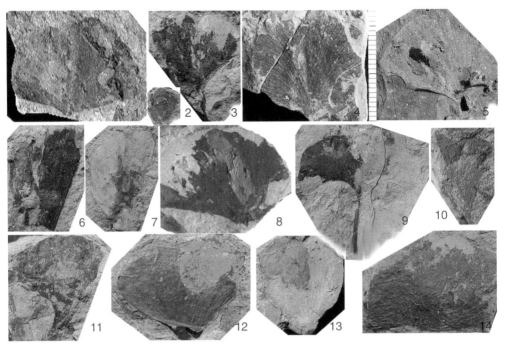

图9 嘉荫晚白垩世银杏类

Fig. 9 Late Cretaceous Ginkgoales in Jiayin

1~8. 似铁线蕨型银杏, CB TP2-191, CB TP2-180, CB TP060920, CB TP0609014, CB TP060915, CB TP205047, CB TP060921, RCPS-2005-45; 9, 10. 具毛银杏, RCPS-BJ-068, CB TP060903; 11~14. 银杏（未定种）CB TP060905, CB TP060928, CB TP060917, CB TP060912

1-8. *Ginkgo adiantoides*, CB TP2-191, CB TP2-180, CB TP060920, CB TP0609014, CB TP060915, CB TP205047, CB TP060921, RCPS-2005-45; 9, 10. *Ginkgo pilifera*, RCPS-BJ-068, CB TP060903; 11-14. *Ginkgo* sp., CB TP060905, CB TP060928, CB TP060917, CB TP060912

　　叶角质层主要特征为：气孔器主要集中见于下表皮，但上表皮偶见气孔；上下表皮均发育大量散布的毛状体。上表皮脉路区通常由4~6列伸长的普通表皮细胞组成，大小为83~121 μm×13~21 μm；垂周壁直或波状弯曲；平周壁不平坦。脉间区普通表皮细胞多呈不规则长方形或多边形，大小为31~56 μm×18~31 μm；垂周壁弯曲；平周壁不平。毛状体尖圆锥状，长可达40 μm，基部直径约16 μm，散布于脉路区与脉间区普通表皮细胞上。偶见气孔。

　　下表皮脉路区由5~7列长矩形或梭形的普通表皮细胞组成；垂周壁直或微弯；平周壁发育强烈的乳突，并可见毛状体。脉间区普通表皮细胞的垂周壁直或微弯；平周壁上可见乳突，并有毛状体散布；毛状体圆锥状，长可达45 μm，基部直径约18 μm（图11，9）。气孔器单环式，大小为19~26 μm×18~23 μm；分布无规则，孔缝无定向；保卫

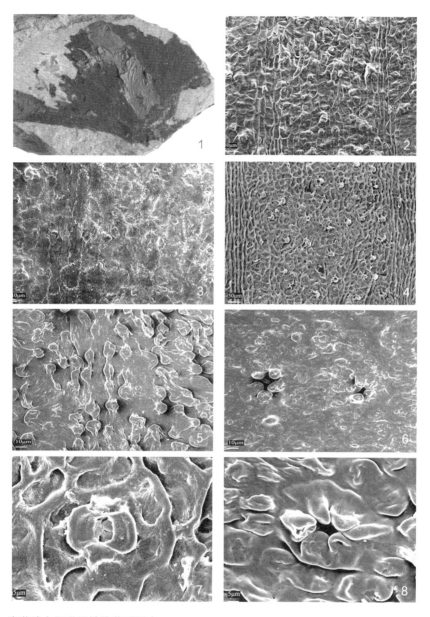

图 10　嘉荫晚白垩世似铁线蕨型银杏

Fig. 10　*Ginkgo adiantoides* of Upper Cretaceous in Jiayin

1. 叶化石,RCPS-2005-45。2、3. 上表皮,内面,普通表皮细胞,SEM1404,SEM1405。4~8.下表皮:4.内面,普通表皮细胞,SEM1718;5、6. 外面,示强烈角质化的表皮细胞及副卫细胞乳突,SEM1724,SEM1613;7.内面,气孔器及保卫细胞角质化,SEM1729;8. 外面,副卫细胞及其乳突,SEM1415

1. Fossil leaf, RCP-2005-45; 2、3. Upper cuticle (inside view) showing ordinary epidermal cells, SEM1404, SEM1405; 4–8, Lower cuticle: 4 (inside) ordinary epidermal cells, SEM1718; 5, 6 (outside) strongly cutinized ordinary epidermal cells and subsidiary cells, SEM1724, SEM1613; 7 (inside) and 8 (outside) showing stomata and cutinizations of subsidiary and guide cells, SEM1729, SEM1415

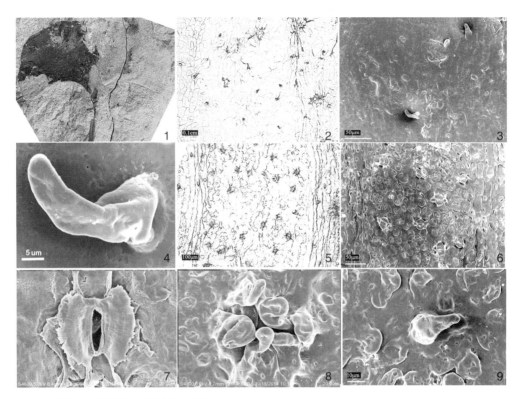

图11 嘉荫晚白垩世具毛银杏表皮构造

Fig. 11　Cuticles of *Ginkgo pilifera* of Upper Cretaceous in Jiayin

1.叶化石,RCPS-BJ-068。2~4.上表皮外面:2、3示普通表皮细胞及毛状体,LM4326,SEM1931;4示毛状体,SEM1811。5~9.下表皮:5、6为内面,示气孔器及毛状体分布,LM4335,SEM1905;7(内面)及8(外面),示气孔器及副卫细胞强烈角质化,SEM0710,SEM0701;9为外面,示毛状体,SEM1907

1. Fossil leaf, RCPS-BJ-068. 2-4. Upper cuticles: 2 and 3（outside）showing ordinary epidermal cells and tricomes, LM4326, SEM1931; 4. Tricomes, SEM1811; 5-9. Lower cuticles: 5 and 6（inside）showing distributions of stomata and tricomes, LM4335, SEM1905; 7（inside）and 8（outside）showing stomata and cuticulazations of subsidiary cells, SEM0710, SEM0701; 9.（outside）showing a tricome, SEM1907

细胞近半圆形,内缘呈放射状角质化加厚(图11,7);副卫细胞5~6个,均具强烈角质化乳突,几掩气孔缝(图11,8)。

当前材料与本种俄罗斯西伯利亚晚白垩世模式标本(Samylina,1967)基本一致,区别仅在于俄罗斯模式材料的上表皮普通表皮细胞具中央加厚或乳突。

产出层位:永安村组(K_{2yn})及太平林场组(K_{2p})。

银杏(未定种) *Ginkgo* sp. (图9,11~14;图12)

叶扇形至半圆形,长1.9~3.1 cm,宽3.2~4.4 cm;全缘微波状或中间浅裂一次,至

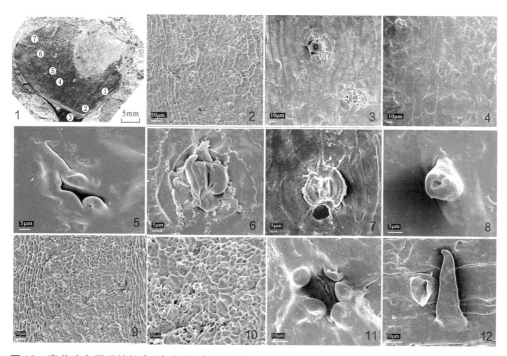

图12　嘉荫晚白垩世的银杏（未定种）表皮构造

Fig. 12　Cuticles of *Ginkgo* sp. from Upper Cretaceous of Jiayin

1. 大化石照片示角质层采集部位，TP060928。2~8. 上表皮：2~4. 叶片基部、内面，普通表皮细胞及气孔分布；5. 外面，示气孔器副卫细胞角质化乳突；6、7. 内面，示气孔器下陷及保卫细胞内缘加厚；8. 外面，示毛状体。9~12. 下表皮：9、10. 内面，示普通表皮细胞及气孔分布；11、12. 外面，11示气孔器副卫细胞角质化乳突；12. 示毛状体。2~12 编号，SEM1627，SEM1741，SEM1702，SEM1729，SEM17302，SEM0705，SEM1743，SEM1645，SEM1646，SEM1807，SEM1805（据公繁浩，2007）

1. Specimen photo showing positions of collecting cuticles, TP060928; 2–8. Upper cuticles: 2–4 basal part of leaf, inside, showing ordinary epidermal cells and stomatal distributions; 5. Outside, showing cutinized thickening of subsidiary cells; 6,7. Inside: showing stomata sunken and subsidiary cells thickening; 8. Outside: trichome; 9–12. Lower cuticles：9, 10. Inside: showing ordinary epidermal cells and stomatal distributions; 11, 12. Outside: 11 showing cutinized thickening of subsidiary cells; 12 showing trichome. SEM1627, SEM1741, SEM1702, SEM1729, SEM17302, SEM0705, SEM1743, SEM1645, SEM1646, SEM1807, SEM1805（after Gong, 2007）

叶片上 1/8~1/4 处；具柄，可见长约 1.2 cm，宽约 1.5 mm；叶基角约 80°~170°；叶脉自基部发出，二歧分叉，每厘米 13~15 条。

表皮构造为双面气孔式（amphistomatal）。上表皮气孔少且分布不均匀，脉路区由 4~7 列长矩形或梭形细胞组成，大小为 65~72 μm×11~16 μm；脉间区多呈四边形或不规则多边形，大小为 50~60 μm×27~42 μm；垂周壁弯曲，有时可见近 Ω 形；平周壁平或微凸，可见乳突或零星散布的毛状体；毛状体钝圆锥状，长可达 20 μm，基部直径可达 13 μm。气孔分布不均匀，主要分布于叶片基部，零星分布于近叶缘，孔缝无定向或局部呈纵向排列。气孔器单唇式，双环式排列，大小为 15~20 μm×18~22 μm；保卫细胞强烈下陷；副卫细胞 4~7 个，角质层加厚或呈乳突状，常不完全掩盖孔口。

下表皮脉路区普通表皮细胞伸长，大小为 48~69 μm×13~28 μm；垂周壁直或微弯；平周壁常加厚或发育乳突，多呈串珠状，直径 8~15 μm，偶见毛。脉间区普通表皮细胞呈不规则四边或多边形，大小为 48~69 μm×13~28 μm；垂周壁直或波状弯曲；平周壁不平坦，普通表皮细胞多发育一个中央乳突，直径 8~10 μm。气孔带宽 156~187 μm，气孔无规则排列，孔缝无定向；气孔器单环式，大小 21~29 μm×19~24 μm；保卫细胞呈新月至半圆形，强烈下陷；副卫细胞 4~6 个，具强烈乳突，几遮掩气孔缝（公繁浩，2007）。

当前材料以典型的双面气孔式区别于 *G. adiantoides*、*G. pilifera* 及 *G. pluripartita* 等白垩纪常见种。与同为双面气孔式的银杏种 *G. huttoni* 和 *G. yimaensis* 等的主要区别在于叶片分裂方式明显不同。原作者认为，当前材料在种级分类上尚待进一步研究（公繁浩，2007）。

产出层位：太平林场组（K$_{2tp}$）。

松柏类 Coniferales

准落羽杉（未定种）*Parataxodium* sp.（图 13，1）

化石保存营养枝，枝轴宽通常约 2 mm。叶线形，互生，羽状排列，与轴的夹角为 30°~40°，叶长 5~15 mm，宽 1~2 mm 居多，顶端尖至钝圆，中脉 1 条，但不十分明显；叶基部略下延于轴。

本属的营养枝叶形态与水杉（*Metasequoia*）和落羽杉（*Taxodium*）均有些相似，但本属的枝轴相对较粗直。

产出层位：永安村组（K$_{2yn}$）及太平林场组（K$_{2tp}$）。

奥尔瑞克落羽杉 *Taxodium olrikii* (Herr) Brown＊（图13,2、3）

张志诚(1984)曾发现本种营养枝及生殖枝。其图片及描述参见该作者原著(张志诚,1984,第120页,图版3,图6、8、10)。叶针状或线披针形,单脉,长约1.5 cm,宽约1 mm,顶端尖锐,基部收缩下延于轴,排成两列状;生殖枝的叶特化,较小,螺旋状排列,具一中脉,球果直径约3 mm,被有鳞片,可能为雄球果。当前新发现的标本(图13,3)与张志诚描述的标本在形态特征上十分接近。

产出层位:永安村组(K_{2yn})及太平林场组(K_{2tp})。

二列水杉 *Metasequoia disticha* (Heer) Miki（图13,4~8）

化石仅保存营养枝部分,长3~5 cm,宽1.5~2.5 cm,枝轴宽1~2 mm。小枝上的叶对生,与枝轴夹角通常为40°~60°。叶线形,扁平,长10~15 mm,宽2~3 mm,顶端钝

图13　嘉荫晚白垩世松柏类(1)

Fig. 13　Late Cretaceous Coniferales of Jiayin（1）

1. 准落羽杉(未定种),RCPS-TP2-1028;2、3. 奥尔瑞克落羽杉(据张志诚,1984),3. RCPS-TP2-1033;4~8. 二列水杉,RCPS-TP2-1026,RCPS-TP2-1027,RCPS-TP2-1028,RCP-TP2-1029,RCP-TP2-1030;9~11. 红杉(未定种),RCPS-TP2-1031,RCPS-TP2-1032;12、13. 水松(未定种),RCPS-TP2-1034,Yn-2-180

1. *Parataxodium* sp., RCPS-TP2-1028; 2, 3. *Taxodium olrikii* (after Zhang, 1984), 3. RCPS-TP2-1033; 4-8. *Metasequoia disticha*, RCPS-TP2-1026, RCPS-TP2-1027, RCPS-TP2-1028, RCP-TP2-1029, RCP-TP2-1030; 9-11. *Sequoia* sp., RCPS-TP2-1031, RCPS-TP2-1032; 12, 13. *Glyptostrobus* sp., RCPS-TP2-1034, Yn-2-180

尖,基部收缩为钝圆形,并具一短柄,中脉清晰且较直。

从外部形态看,本种最明显的特征是:叶羽状对生,且基部收缩具一短柄。当前化石标本的枝叶形态特征与本种的典型标本(Heer,1878,第33页,图版8,图25b;图版9,图1;第52页,图版15,图10~12)基本一致。

产出层位:永安村组(K_{2yn})及太平林场组(K_{2tp})。

红杉(未定种) Sequoia sp. (图13,9~11)

化石保存营养枝,长2~5 cm,宽0.7~1.5 cm,轴宽约1.5 mm。叶全缘线形至椭圆形,顶端较钝,长5~15 mm,宽1~3 mm;叶螺旋状排列,与枝轴夹角约45°,基部不收缩,沿枝下延,排列稀疏,中脉不明显。

当前化石叶螺旋状排列及叶基部沿枝下延等为其较突出的特征。

产出层位:永安村组(K_{2yn})及太平林场组(K_{2tp})。

水松(未定种) Glyptostrobus sp. (图13,12、13;图14,13)

营养枝长3~11 cm,宽1~1.6 cm;枝轴较细,宽1~1.5 mm。叶线形,长4~10 mm,宽约1 mm;互生,与枝轴夹角较小,为15°~30°;叶顶端尖,基部下延至轴,中脉不清晰。其球果鳞片零散保存,单个种鳞呈扇形,长约5.0 mm,宽约3 mm。

产出层位:永安村组(K_{2yn})及太平林场组(K_{2tp})。

落叶松(未定种) Larix sp. (图14,1)

叶线性,近针状,见7~8枚放射状簇生于一短枝上。叶长8~12 mm,宽0.5~0.7 mm,似具一中脉,但不清晰。短枝长3~7 mm,宽约4 mm。未保存生殖部分。

当前标本在叶及短枝等形态特征上与现生的一些落叶松的种(如 Larix sibirica Led.)较为相似。因当前标本保存不完整,又缺乏球果等生殖部分,暂难做进一步比较。

产出层位:太平林场组(K_{2tp})。

小松型籽 Pityospermum minutum Samylina (图14,2)

翅果倒卵形,长约7 mm,向上渐窄,顶端钝圆,翅表面具纵纹。种子生于翅基部,卵圆形,半径约3 mm,翅呈长三角形。

产出层位:永安村组(K_{2yn})及太平林场组(K_{2tp})。

司氏柏型枝 Cupressinocladus sveshnikovae Ablaev (图14,4~8)

化石保存营养枝。枝轴宽3~5 mm,小枝互生。叶鳞片形,螺旋状着生于小枝上;叶长5~15 mm,宽2~3 mm,顶端多钝圆。

当前标本与俄罗斯南滨海本种模式标本(Ablaev,1974)以及孙革等描述的嘉荫

图14 嘉荫晚白垩世松柏类(2)

Fig. 14 Late Cretaceous Coniferales of Jiayin（2）

1. 落叶松(未定种), RCPS-TP2-105; 2. 小松型籽, RCPS-TP2-1036; 3. 松型叶(未定种), RCPS-TP2-1037; 4~8. 司氏柏型枝, RCPS-TP2-1038, RCPS-TP2-1081, RCPS-TP2-1082, RCPS-TP2-1083, RCPS-TP2-1084; 9. 枞型枝(未定种1), RCPS-TP2-1039; 10. 11. 枞型枝(未定种2), RCPS-TP2-1040, RCPS-TP2-1041; 12. 石籽(未定种), RCPS-TP2-020; 13. 水松(未定种), 球果鳞片, RCPS-TP2-1042

1. *Larix* sp., RCPS-TP2-105; 2. *Pityospermum minutum*, RCPS-TP2-1036; 3. *Pityophyllum* sp., RCPS-TP2-1037; 4-8. *Cupressinocladus sveshnikovae*, RCPS-TP2-1038, RCPS-TP2-1081, RCPS-TP2-1082, RCPS-TP2-1083, RCPS-TP2-1084; 9. *Elatocladus* sp. 1, RCPS-TP2-1039; 10, 11. *Elatocladus* sp. 2, RCPS-TP2-1040, RCPS-TP2-1041; 12. *Carpolithus* sp., RCPS-TP2-020; 13. *Gryptostrobus* sp., synphyllodium, RCPS-TP2-1042

同种标本(Sun et al., 2007, 2011)在形态特征上基本一致。

产出层位:永安村组(K_{2yn})及太平林场组(K_{2tp})。

枞型枝(未定种1) *Elatocladus* sp. 1(图14,9)

化石保存营养枝。枝轴较细,宽约1 mm,可见长约2 cm,宽约1.5 cm;小枝长近1 cm,宽1.5~2 mm,近宽线形,小枝轴宽约0.5 mm;叶近对生,近宽镰形,略上弯,每枚叶长1~1.5 mm,基部最宽处0.5~0.7 mm;顶部渐尖、近刺状,似有一条中脉,叶基部紧贴于小枝着生。

当前标本的具叶枝条形态与Sveshnikova（1963)报道的产于我国云南的现生秃杉(*Taiwania flousiana* Gaussen)的异形叶的枝条(Sveshnikova, 1963, 第219页, 图版

15,图5)以及她同时报道的乌克兰西南始新世的台湾杉(*Taiwania*)的叶子(Sveshnikova,1963,图版15,图12等)均有些相似。由于当前标本缺少生殖部分保存,暂难做详细分类;但不排除当前标本可能属于晚白垩世杉科的一个新的分类群,有待进一步研究。

产出层位:太平林场组(K_{2tp})。

枞型枝(未定种2) *Elatocladus* **sp. 2**（图14,10,11）

化石保存为一块带叶营养小枝。枝纤细,直径约0.8 mm。叶线形,近对生于枝上。叶长7~12 mm,宽约1 mm;叶片直或中上部略向轴一侧弯曲,基部微收缩并下延,与轴呈约45°角;中脉粗直。

产出层位:永安村组(K_{2yn})及太平林场组(K_{2tp})。

石籽(未定种) *Carpolithus* **sp.**（图14,12）

种子化石,近球形,长径约7 mm,外种皮似薄壳状,表面光滑。化石零散保存且不完整。分类暂时不明。

产出层位:太平林场组(K_{2tp})。

查加扬罗汉松(比较属种) **cf.** *Podocarpus tsagajanicus* **Krassilov**

叶宽线至披针形,长3.4 cm,宽3~4 mm,顶端钝尖,向下缓缓变窄,具单脉。本种详细特征可参见张志诚(1984)描述。

产出层位:太平林场组(K_{2tp})。

松型叶(未定种) *Pityophyllum* **sp.**

叶线性,长2.5 cm,宽约3 mm,顶端钝尖,具单脉,基部未保存。

产出层位:太平林场组(K_{2tp})。

被子植物 Angiospermae

惚木叶?(未定种) *Araliaephyllum?* **sp.**（图15,1）

单叶,全缘,叶顶部似分裂,基部楔形具一短柄,叶脉羽状,主脉直,二级脉对生排列,弧状弯曲向上直达叶边缘,三级脉不清晰,化石保存不完整。

产出层位:永安村组(K_{2yn})及太平林场组(K_{2tp})。

车氏阿朔叶 *Arthollia tschernyschewii* **(Konstantov) Golovneva, Sun et Bugdaeva**（图15,2;图18,1~3）

单叶,全缘,卵圆形至椭圆形,叶基部楔形或收缩为心形,常具一短柄,叶顶部渐尖,长5~15 cm,宽3~10 cm;叶边缘具细齿且微凹,叶脉羽状,直行脉序,中脉直,二级

图 15　嘉荫晚白垩世被子植物化石（1）

Fig. 15　Late Cretaceous angiosperms of Jiayin（1）

1. 惚木叶?（未定种）；2. 车氏阿朔叶，RCPS-TP2-2002-122；3. 东方阿朔叶，MH1085；4~8. 昆都尔似
南蛇藤，RCPS-TP2-2002-113，RCPS-TP2-2002-155，RCPS-TP2-2002-114，RCPS-TP2-2002-301，
RCPS-TP2-2002-302；9. 密脉悬铃木，MH1064；10、11. 中华悬铃木，MH1069，MH1071；12. 悬铃木
（未定种），Yn-B-016（3，9~11 据张志诚，1984）

1. *Araliaephyllum?* sp.; 2. *Arthollia tschernyschewii*, RCPS- TP2- 2002- 122; 3. *Arthollia orientalis*,
MH1085; 4-8. *Celastrinites kundurensis*, RCPS-TP2-2002-113, RCPS-TP2-2002-155, RCPS-TP2-
2002-114, RCPS-TP2-2002-301, RCPS-TP2-2002-302; 9. *Platanus multinervis*, MH1064; 10, 11.
Platanus sinensis, MH1069, MH1071; 12. *Platanus* sp., Yn-B-016（3, 9-11, after Zhang, 1984）

脉7~8对,互生,三级脉呈梯纹状,相互吻合连接排列。

当前标本叶形体小且基部心形区别于 *A. pacifica* 和 *A. inordinata*。此外,与 *A. pacifica* 相比,当前种二级脉在叶边缘的分叉次数明显较多。

产出层位:太平林场组(K_{2tp})。

东方阿朔叶 *Arthollia orientalis* (Zhang) Golovneva（图15,3）

单叶,全缘,宽卵形,下部宽8~9 cm,长7~8 cm,基部楔形或心形,顶端钝圆。掌羽状脉序,中脉直,达顶,侧主脉发达,二级脉稍弯曲与主脉的夹角为30°~50°,三级脉形成梯纹状吻合结构。本种原定名为东方似翅子(*Pterosperites orientalis* Zhang)(张志诚,1984),后属名被高洛夫涅娃修订为阿朔叶(*Arthollia*);本种修订后的新描述可参见 Golovneva et al.（2008）。

产出层位:太平林场组(K_{2tp})。

昆都尔似南蛇藤 *Celastrinites kundurensis* Golovneva,Sun et Bugdaeva（图15,4~8）

单叶,具柄,叶长椭圆形或披针形,长3~18 cm,宽1~6 cm,叶顶部圆形或微尖,基部楔形或截形,常不对称,叶边缘,上半部具小齿或锯齿,下半部全缘;叶脉羽状环结脉序,中脉直,较粗,近基部直径可达4 mm;二级脉8~11对,互生或近对生,直或微弧状,与中脉的夹角为40°~50°,平行伸展至叶边缘处形成环状,三级脉梯纹状吻合交接,形成多边形的网格。

当前标本与北美古新世 *C. insignis* 相比,后者叶的形体更大(长达20 cm)、叶缘自基部到顶部均具齿、二级脉更多(10~15对)且夹角更大(50°~80°)。*C. septentrionalis* 虽与本种均为狭长叶形,但其叶的形态变化较多,且叶基部最宽,此外其层位往往偏新。

产出层位:太平林场组(K_{2tp})。

密脉悬铃木 *Platanus densinervis* Zhang（图15,9）

大型叶,宽椭圆形,可见长大于5 cm,中下部最宽处宽约3 cm;近全缘。中脉直,较细,侧脉细,自中脉呈羽状分出,与中脉夹角为60°~70°;三级脉呈网状,但不很清楚。生殖部分未保存。

产出层位:太平林场组(K_{2tp})。

中华悬铃木 *Platanus sinensis* Zhang（图15,10、11）

叶形变化较大,小叶近圆形,上部不分裂或三浅裂,下部圆楔形;大叶呈三裂状,裂片顶端钝圆或钝尖,边缘具齿。掌状三出脉,较细,达叶边缘,中脉直,二级脉3~5

对,对生或近对生,三级脉以宽角伸出,结成长方形网格。详见张志诚(1984)。

产出层位:太平林场组(K_{2tp})。

悬铃木(未定种) *Platanus* **sp.**（图15,12）

单叶,扁圆形,宽4.6 cm,高3.6 cm,上部三浅裂,下部平截,基部微下延,叶边缘具浅凹齿。掌状三出脉,中脉直达顶,侧主脉以45°自基部伸出后微弯曲达两侧裂片顶端,二级脉羽状,互生,三级脉不清。详见张志诚(1984)。

产出层位:永安村组(K_{2yn})太平林场组(K_{2tp})。

嘉荫达勒比叶 *Dalembia jiayinensis* **Sun et Golovneva**（图16）

奇数羽状复叶,具5枚小叶。小叶呈椭圆形、卵形或近三角形,顶端钝圆,小叶基部呈楔形、截形或微心形,略不对称;叶缘全缘或裂缺。脉序为羽状或掌状羽状,侧脉直行或半直行。

本属以往仅在俄罗斯远东地区发现且层位相对较低(赛诺曼期—康尼亚克期,Cenomanian–Coniacian),此次新发现首次扩大了对该属时代及古地理分布的认识,即

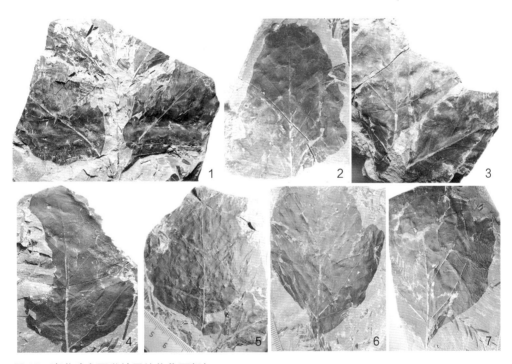

图16　嘉荫晚白垩世被子植物化石(2)

Fig. 16　Late Cretaceous angiosperms of Jiayin (2)

1~7.嘉荫达勒比叶

1–7. *Dalembia jiayinensis*

该属所属时代可达到晚白垩世桑顿期（Santonian），分布范围最南端可达中国黑龙江嘉荫。

产出层位：永安村组（K_{2yn}）。

北极准紫树果 *Nyssidium arcticum* (Heer) Iljinskaja（图17,1）

总状果序化石，长卵形，长4~8 cm，宽2~3 cm，由4~16个螺旋状排列单果组成，每个单果具一长约1 mm的短柄；单果卵圆形或椭圆形，长5~8 mm，宽4 mm，中部宽向两端变尖，表面具细的平行的长条状纹饰。当前标本与冯广平等报道的产于嘉荫乌云组的 *Nyssidium jiayinense* Feng et al. 有些相似，但后者单果数量较多，且内果皮表面具横向纤维（横纹）（Feng et al., 2000）。

产出层位：永安村组（K_{2yn}）。

北极似昆栏树 *Trochodendroides arctica* (Heer) Berry（图17,2~4；图18,6）

单叶，长卵形、椭圆形或卵形，叶形变化较大，具长叶柄，大小不一，叶顶部钝圆，基部楔形、圆楔形或心形，叶边缘具较大的圆齿或波状；掌状三出脉为本属典型特征，个别较宽叶片为掌状五出脉，中脉直达叶顶，侧脉弧状弯曲，三级脉羽状分支，常结成梯纹状网格。

当前标本叶脉特征与现生连香树 *Cercidiphyllum* 属较为相似，但前者果实常为单果，总状果序。

产出层位：永安村组（K_{2yn}）及太平林场组（K_{2tp}）。

披针形似昆栏树 *Trochodendroides lanceolata* Golovneva, Sun et Bugdaeva（图17,5~10；图18,5）

单叶，多为披针形或长卵形，为本种典型特征，其叶片上部略膨大，叶尖钝圆，边缘具圆齿或锯齿，叶下部至基部变窄，呈楔形，边缘全缘，基部不对称；叶脉为掌状脉，环结脉序。当前标本的叶形为披针形或长卵形，区别于本属其他种。

产出层位：太平林场组（K_{2tp}）。

太平林场似昆栏树 *Trochodendroides taipinglinchanica* Golovneva, Sun et Bugdaeva（图17,11~13）

单叶，革质，圆形至宽椭圆形，长1~7 cm，宽0.8~6 cm，具短柄，较粗；叶顶部钝圆，基部宽楔形，叶片上部具规则、直径3~4 mm的小圆齿，掌状环结脉，中脉直，两个侧脉自叶柄基部伸出达叶边缘，并形成环脉，三级脉梯纹、网状，较密集。叶片顶部具规则的小圆齿、下部宽楔形、具短柄是本种主要特征。

产出层位：太平林场组（K_{2tp}）。

图17 嘉荫晚白垩世被子植物化石（3）

Fig. 17 Late Cretaceous angiosperms of Jiayin（3）

1.北极准紫树果，Yn-B-320；2~4.北极似昆栏树，RCPS-TP2-010，RCPS-TP2-303，RCPS-TP2-304；5~10.披针形似昆栏树，RCPS-TP2-2002-31，RCPS-TP2-2002-305，RCPS-TP2-2002-306，RCPS-TP2-2002-307，RCPS-TP2-2002-308，RCPS-TP2-2002-309；11~13.太平林场似昆栏树，RCPS-TP2-2004-2，RCPS-TP2-26，RCPS-TP2-310；14.微尖似昆栏树，RCPS-TP2-011

1. *Nyssidium arcticum*，Yn-B-320；2-4. *Trochodendroides arctica*，RCPS-TP2-010，RCPS-TP2-303，RCPS-TP2-304；5-10. *T. lanceolata*，RCPS-TP2-2002-31，RCPS-TP2-2002-305，RCPS-TP2-2002-306，RCPS-TP2-2002-307，RCPS-TP2-2002-308，RCPS-TP2-2002-309；11-13. *T. taipinglinchanica*，RCPS-TP2-2004-2，RCPS-TP2-26，RCPS-TP2-310；14. *T. microdentatus*，RCPS-TP2-011

微尖似昆栏树 *Trochodendroides microdentatus* (Newb.) Krysht.（图 17，14；图 18，7）

单叶，宽卵形，长 3~10 cm；宽 2.5~5 cm；叶顶急尖，具钝尖叶齿，齿腋钝圆。掌状三出脉，中间一级脉直，向上变细；两侧一级脉较弱，近叶缘区常分叉 2~3 次，各分叉与三级脉结为多边形；三级脉排列呈不规则网格。当前标本叶顶尖头等特征与 Kryshtofovich（1966）描述的本种模式标本十分接近；与 Krassilov（1976）报道的俄罗斯结雅–布列亚盆地标本以及张志诚（1984）报道的嘉荫同位层本种标本均较相似。

产出层位：永安村组（K_{2yn}）及太平林场组（K_{2tp}）。

拟蝙蝠叶（未定种）*Menispermites* sp.

单叶，盾状或肾形，宽大于长，长 4~6 cm，宽 5~7 cm；顶端钝尖，基部收缩，宽心形，具叶柄。上部叶缘具波状圆齿，基部近于全缘。掌状 5~6 基出脉，中脉直，侧脉弯曲，分叉 2~3 次达叶边缘，三级脉较弱，交织成不规则细网格。

产出层位：永安村组（K_{2yn}）。

扭曲荚蒾（比较种）*Viburnum* cf. *contortum* Lesquereux＊（图 18，8）

单叶，卵形，不分裂，长 9 cm，宽 7.5 cm，顶端钝尖，基部宽楔形，叶上部边缘波状，下部全缘。羽状脉序，中脉直，达顶，二级脉 7~8 对，近互生，三级脉结成长方形网格。详见张志诚（1984）对本种的描述（张志诚，1984，第 126 页，图版 7，图 1）。

产出层位：太平林场组（K_{2tp}）。

荚蒾叶（未定种）*Viburnophyllum* sp.（图 18，9）

单叶，长椭圆形，可见长大于 4 cm，宽大于 3 cm，基部宽楔形，叶边缘具细齿。羽状脉序，中脉直，宽约 1 mm，二级脉 4~5 对，略直或微弯，近对生或略互生，以约 30° 自中脉伸出；叶边缘三级脉清晰，略上弯达叶边缘；叶内三级脉较细，近直角自二级脉伸出，与相邻三级脉通常织成近矩形的网格。

产出层位：太平林场组（K_{2tp}）。

北方诺登斯基果（比较种）*Nordenskioideia* cf. *borealis* Heer＊

果实化石，近圆形，直径约 7 mm，约 6 个裂瓣紧密结合，使果实呈一轮状（详见张志诚，1984，第 127 页，图版 3，图 4a）。

产出层位：太平林场组（K_{2tp}）。

图18　嘉荫晚白垩世被子植物化石（4）

Fig. 18　Late Cretaceous angiosperms of Jiayin（4）

1~3. 车氏阿朔叶，TP2-101，TP2-102，TP2-103；4. 双子叶被子植物叶（未定种），RCPS-yn-2-077；
5. 披针形似昆栏树，RCPS-TP2-36；6. 北极似昆栏树，RCPS-TP2-015；7. 微尖似昆栏树，RCPS-TP2-
21；8. 扭曲荚蒾（比较种）；9. 荚蒾叶（未定种），TP2-104（8据张志诚，1984）

1-3. *Arthollia tschernyschewii*，TP2-101，TP2-102，TP2-103；4. *Dicotyleophyllum* sp.，RCPS-yn-2-077；
5. *Trochodendroides lanceolata*，RCPS-TP2-36；6. *T. arctica*，RCPS-TP2-015；7. *T. microdentatus*，
RCPS-TP2-21；8. *Viburnum* cf. *contortum*；9. *Viburnophyllum* sp.，TP2-104（8 after Zhang，1984）

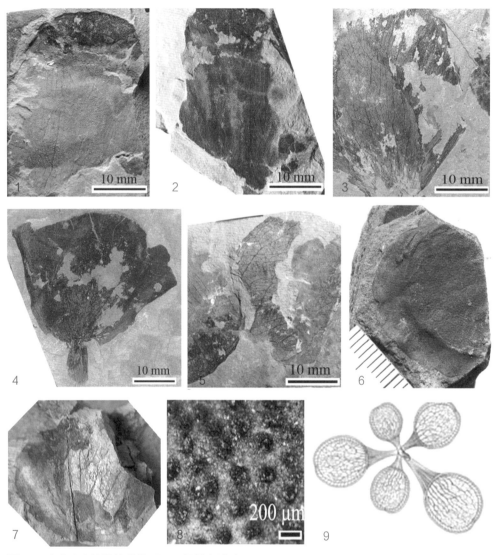

图19　嘉荫水生被子植物化石——褶皱卡波叶

Fig. 19　Aquatic angiosperms of Jiayin：*Cobbania corrugata*

1~4，8. 产自永安村组，Yn-001，Yn-002，Yn-003，Yn-008；5~7. 产自太平林场组，TP2-106，TP2-107，TP2-108；8. 示7的叶上表面表皮细胞；9. 叶的组合复原图（据Stockey et al.，2007）

1-4, 8. From Yong'ancun Fm., Yn-001, Yn-002, Yn-003, Yn-008; 5-7. From Taipinglinchang Fm., TP2-106, TP2-107, TP2-108; 8. Showing the ordinary epidermal cells of upper cuticle from 7; 9. Reproduced picture of leaf combination of *Cobbania corrugate*（after Stockey et al., 2007）

水生被子植物

褶皱卡波叶 *Cobbania corrugata* (Lesq.) Stockey, Ruthwell et Johnson（图19）

水生草本植物，莲座状复叶。叶近圆形或椭圆形，全缘略呈波状或具齿，长1~9 cm，宽1~6 cm。叶的近轴面，叶脉弯曲、细弱，叶缘有较规则的近圆形缘脉，叶面一级脉、二级脉及三级脉交织成网状，形成不规则的多边形细网格，网格内密布小孔，疑似毛状基，直径为50~150 μm；叶的远轴面，一级脉可达7.6 cm，二级脉较弱，并与三级脉交织成不规则网格。叶柄较粗，直径约5~9 mm。未见根部及生殖部分保存。

当前标本在叶部形态特征上与Stockey等（Stockey et al., 2007）描述的本种典型标本的特征基本一致。

产出层位：永安村组（K_{2yn}）及太平林场组（K_{2tp}）。

嘉荫莲 *Nelumbo jiayinensis* Liang, Sun et Yang（图20）

水生草本植物。叶近盾圆形，全缘或略显波状叶缘。叶柄位于叶片中心，具一级脉20~25条，自叶片中心辐射状向叶边缘伸展，在近叶缘处分叉1~2次；二级脉弯曲并交织成网状；三级脉变弱并织成细网格。超景深显微观察见到叶表皮细胞形态：上表皮普通细胞呈较规则的多边形，大小为20~50 μm，垂周壁较直；下表皮普通细胞与上表皮的形态相似，大小为40 μm×30 μm，未见气孔器。

产出层位：主要产于永安村组（K_{2yn}），个别见于太平林场组（K_{2tp}）。

具棱葛赫叶 *Quereuxia angulata* (Newb.) Kryshtofovich（图21）

水生草本植物，小型叶。叶异型，分沉水叶和浮水叶。浮水叶宽卵形至窄倒卵形，叶边缘常具齿，叶柄细长，脉序羽状。沉水叶线形，呈放射状，假二歧分叉多次。Samylina（1988）曾深入研究本属的生态特征并提出复原图（图21，11）。

叶表皮细胞结构呈单面气孔式。上表皮的内表面，普通表皮细胞多呈长方形（50~70 μm×30 μm）或规则的多边形（4~6边），垂周壁略平直，具少量气孔器；气孔器长椭圆形，定向排列，副卫细胞4~6个；外表面普通表皮细胞多呈纵向褶皱状，气孔器定向排列，保卫细胞外侧可见有纵向褶皱状角质化。下表皮普通表皮细胞多呈多边形（4~6边），未见气孔器。

产出层位：永安村组（K_{2yn}）及太平林场组（K_{2tp}）。

2.1.2　孢粉化石

黑龙江嘉荫晚白垩世植物群中，在发现植物大化石的同时，还发现有丰富的孢粉化石，这为研究嘉荫晚白垩世植物群的组成及时代等发挥了重要作用。

黑龙江嘉荫晚白垩世植物群

图20　嘉荫水生被子植物化石——嘉荫莲

Fig. 20　Aquatic angiospermae of Jiayin: *Nelumbo jiayinensis*

1~9. 嘉荫莲, 1为模式标本, YX-B-302; 2为叶柄部放大; 4为3的叶脉局部放大, YX-B-301; 7为5及8的叶脉间表皮细胞形态, YX-B-308; 9为叶下部化石, TP2-100; 1~8产自永安村组; 9产自太平林场组。10. 嘉荫莲古生态环境复原示意(王文帅绘制, 2019)

1-9. *Nelumbo jiayinensis*; 1. Holotype, YX-B-302; 2 showing petiole enlarged; 3-9. Leaves, 4 showing the veins of 3 in detail, YX-B-301; 7 showing the epidermal structure of 5 and 8, YX-B-308; 9 showing a lower part of leaf, TP2-100; 1-8 from Yong'ancun Fm., 9 from Taipinglinchang Fm.; 10. paleoecological reconstruction of *Nelumbo jiayinensis* (drawn by Wang W. S., 2019)

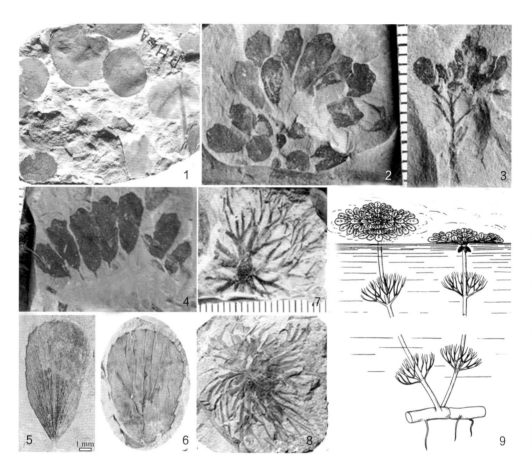

图21 嘉荫水生被子植物化石——具棱葛赫叶

Fig. 21 Aquatic angiosperms of Jiayin: *Quereuxia angulata*

1~6. 浮水叶, Yn-021, Yn-022, Yn-023, P1H42A, Yn-024, Yn-025; 7、8. 沉水叶, Yn-026, Yn-027; 9. 古生态复原图(据Samylina, 1988)

1~6. Floating leaves, Yn-021, Yn-022, Yn-023, P1H42A, Yn-024, Yn-025; 7, 8. Submerged leaves, Yn-026, Yn-027; 9. Reconstruction of paleoecology (after Samylina, 1988)

嘉荫晚白垩世孢粉化石可划分为5个孢粉组合,自老至新分别为:①三塔库波里粉-北方双型孢组合(以永安村组的孢粉为代表,时代为桑顿期);②三棱鹰粉-多皱罗汉松粉组合(以太平林场组为代表,坎潘期);③阿米格黛鹰粉-明确买麻藤粉组合(以渔亮子组下段为代表,马斯特里赫特早期);④粗糙沃氏粉-黑龙江小突粉组合(以渔亮子组上段为代表,马斯特里赫特中期);以及⑤三棱鹰粉-条纹假鹰粉组合(以富饶组为代表,马斯特里赫特晚期)。

① *Kuprianipollis santaloides-Duplosporis borealis* (三塔库波里粉-北方双型孢)组合(组合 I)(图22,A)

本组合主要以嘉荫永安村组孢粉化石为代表,以裸子植物的松科、杉科的花粉以及银杏苏铁粉(*Ginkgocycadophytus*)为主;蕨类孢子以蕨科、桫椤科和蚌壳蕨科为主;被子植物花粉以胡桃科、山毛榉科和悬铃木科为主,库波里粉(*Kuparianipollis*)最为常见。代表性分子主要包括*Cicatricosisporites dorogensis* Pot. et Gell.,*Rouseisporites reticulatus* Chlon.,*Cyathidites minor* Coup.,*Classopollis classoides* Pfl. emend Poc. et Jans.,*Laevigatosporites ovoideus* Takah.,*Taxodiumpollenites hiatus* Pot. et Kremp.,*Kuprianipollis elegans* (Zakl.) Kom.,*K. santaloides* (Zakl.) Kom.,*Cetripites cretaceus* Poc. 等;细弱鹰粉(*Aquilapollenites subtilis*)也常见。本组合在结雅-布列亚盆地显示的时代为桑顿期,在嘉荫显示的时代也被认为是桑顿期(Markevich et al.,2006,2011;孙革等,2014)。

② *Aquilapollenites conatus-Podocarpidites multesimus* (三棱鹰粉-多皱罗汉松粉)组合 (组合 II)(图22,B)

本组合主要以嘉荫太平林场组孢粉化石为代表。本组合蕨类孢子以桫椤科和水龙骨科为代表(达25%);裸子植物花粉以松科和杉科为主,买麻藤科、掌鳞杉科和银杏类花粉较少见;被子植物花粉分异度较高(可达30个分类群),主要以悬铃木科、山毛榉科、榆科、胡桃科等为代表,鹰粉(*Aquilapollenites*)不多(仅占4%~5%)。本组合常见分子主要有*Pterisporites cretacea*,*Podocarpidites ellipticus*,*P. multesimus*,*Polycingulatus densatus*,*Aquilapollenites conatus*,*A. rombicus*,*A. amplus* 等。显示时代为坎潘期(Markevich et al.,2006,2011;孙革等,2014)。

③ *Aquilapollenites amygdaloides-Gnetaceaepollenites evidens* (阿米格黛鹰粉-明确买麻藤粉)组合(组合III)(图22,C)

本组合主要以嘉荫龙骨山渔亮子组下段含恐龙层的孢粉化石为代表。其裸子植物花粉以松科和杉科占优势,但银杏粉及买麻藤粉有所增加。被子植物花粉中,三沟粉占优势,榆科、胡桃科、桦科及山毛榉科等常见。本组合代表性分子主要包括*Tri-*

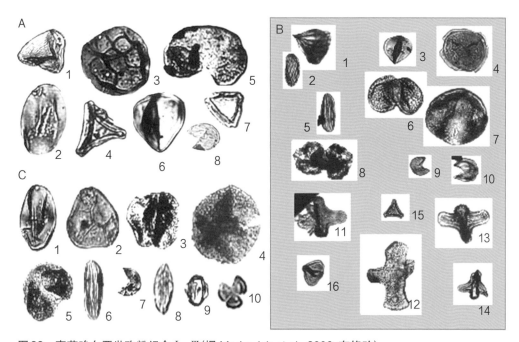

图22　嘉荫晚白垩世孢粉组合Ⅰ～Ⅲ(据 Markevich et al.,2006,有修改)

Fig. 22 Late Cretaceous palynological assemblages Ⅰ－Ⅲ in Jiayin (after Markevich et al., 2006,with revision)

A. 三塔库波里粉–北方双型孢组合(组合Ⅰ) *Kuprianipollis santaloides–Duplosporis borealis* Ass. (Ass. Ⅰ): 1. *Cicatricosisporites dorogensis*; 2. *Laevigatosporites ovoideus*; 3. *Rouseisporites reticulatus*; 4. *Kuprianipollis elegans*; 5. *Cetripites cretaceus*; 6. *Cyathidites minor*; 7. *Kuprianipollis santaloides*; 8. *Taxodiumpollenites hiatus*

B. 三棱鹰粉–多皱罗汉松粉组合(组合Ⅱ) *Aquilapollenites conatus–Podocarpidites multesimus* Ass. (Ass. Ⅱ): 1. *Appendicisporites cristatus* (Mark.) Poc.; 2. *Gnetaceaepollenites ovatus* (Pierce) Verb.; 3. *Cyathidites minor* Coup.; 4. *Rouseisporites reticulatus* Chlon.; 5. *Gnetaceaepollenites evidens* (Bolch.) Chlon.; 6. *Podocarpidites multesimus* (Bolch.) Pock.; 7. *Alisporites bilateralis* Rouse; 8. *Podocarpidites multesimus* (Bolch.) Pock. 9,10. *Taxodiumpollenites distichiforme* (Simps.) Sriv.; 11. *Aquilapollenites insignis* N. Mtch.; 12. *Kuprianipollis elegans* (Zakl.) Kom.; 13. *Aquilapollenites dispositus* (N. Mtch.) Funkh.; 14. *Cranwellis striata* (Coup.) Sriv.; 15. *Aquilapollenites conatus* Nort.; 16. *Aquilapollenites gracilis* Markev

C. 阿米格黛鹰粉–明确买麻藤粉组合(组合Ⅲ) *Aquilapollenites amygdaloides–Gnetaceaepollenites evidens* Ass. (Ass. Ⅲ): 1. *Laevigatosporites ovoideus*; 2. *Rouseisporites reticulatus*; 3. *Podocarpites multissimus*; 4. *Osmandatites welmanii*; 5. *Cedruspollinites perassiccmus*; 6. *Gnetaleapolliinites evidens*; 7. *Taxodiumpollenites hiatus*; 8. *Ginkgocydophytus* sp.; 9. *Orbisuspollis incidus*; 10. *Tricolpites reticulatus*

colpites mataurensis，*T. reticulatus*，*Retitrescolpites concinnatus*，*Kurtzipites tricuspidatus* 等。这一组合显示的时代为马斯特里赫特早期(early Maastrichtian)(Markevich et al.，2006，2011；孙革等，2014)。

④ *Wodehouseia aspera-Parviprojectus amurensis* (粗糙沃氏粉-黑龙江小突粉)组合(组合Ⅳ)(图23)

本组合主要以嘉荫乌拉嘎渔亮子组上段含恐龙层的孢粉化石为代表。其裸子植物花粉主要为松科、杉科和银杏粉；蕨类孢子中 *Laevigatosporites* 和 *Cyathidites* 较多；被子植物花粉可达44%，多样性增加，以桦科、胡桃科、山毛榉科、榆科和杨梅科(Myricaceae)等为主，似代表适度温和的气候。本组合代表性分子主要包括 *Proteacidites mollis*，*Aquilapollenites rigidus*，*Parviprojectus amurensis*，*Wodehouseia aspera*，*W. gracilis* 等。本组合显示时代为马斯特里赫特中期 (middle Maastrichtian)(Markevich et al.，2009，2011；孙革等，2014)。

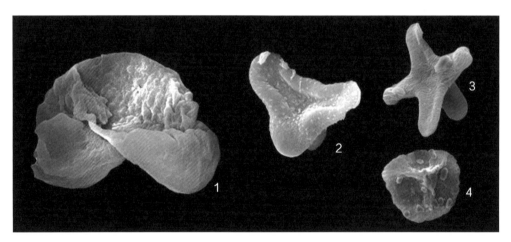

图23　粗糙沃氏粉-黑龙江小突粉组合(组合Ⅳ)(据Markevich et al.，2006)
Fig. 23　*Wodehouseia aspera-Parviprojectus amurensis* Assemblage (Ass. Ⅳ)(after Markevich et al.，2006)

1. *Cedripites parvisaccatus* (Zauer) Chlon.；2. *Aquilapollenites insignis* N. Mtch.；3. *Aquilapollenites striatus* Nort.；4. *Nevesisporites radiatus* Chlon.

本组合的建立主要参考了与嘉荫邻区、俄罗斯结雅-布列亚盆地马斯特里赫特中期孢粉化石组合的对比。本组合的建立为研究嘉荫地区晚白垩世植物群的演化、恐龙灭绝的时代，以及嘉荫晚白垩世地层划分对比等，发挥了重要作用(Sun et al.，2011；孙革等，2014)。

⑤ *Aquilapollenites conatus-Pseudoaquilapollenites striatus*（三棱鹰粉-条纹假鹰粉）组合（组合Ⅴ）（图24）

本组合主要以嘉荫富饶组的孢粉化石为代表，迄今已发现至少29属49种；样品主要采自嘉荫乌云小河沿附近的三个钻孔（XHY-2005，XHY-2006，XHY-2008）富饶组的岩心。本组合孢粉化石主要以裸子植物的杉科（约占50%）和被子植物的榆科（占25%）占优势。孢子中以单缝的 *Laevigatosporites* 占优势，*Cyathidites minor*、*C. australis* 和 *Leiotrileles* spp. 等居第二位。被子植物具有较丰富的多样性，其花粉以榆科、桦科、杨柳科、胡桃科、杨梅科和桃金娘科（Myrtaceae）等为主；鹰粉的分异度达到最高，已发现至少14种，包括 *Aquilapollenites attenuatus* Funkh.，*A. rigidus* Nort.，*A. spinulosus* Funkh.，*A. stelkii* Sriv.，*A. catenireticulatus* Sriv.，*A. striatus* Funkh.，*A. conatus* Nort.，*A. insignis* N. Mtch.，*A. subtilis* N. Mtch.，*A. reductus* Nort.，*A. proceros* Samoil.，*A. amurensis*（Bratz.），*A. quadricretaceus* N. Mtch.，*A. rombicus*（Samoil.）Stanl. 等。此外，近年来俄罗斯孢粉学者 Tekleva 等对产自嘉荫小河沿富饶组的被子植物花粉 *Wodehouseia spinata*，*Pseudointegricorpus clarireticulatum* 以及 *Aquilapollenites* 的三个种进行了花粉壁超微结构深入研究，取得重要进展（Tekleva et al.，2015，2019，2020）。

本组合最典型的代表性分子有8种，分别为：*Marsypiletes cretacea*，*Tricolpites variexinus*，*Aquilapollenites stelkii*，*A. proceros*，*A. striatus*，*A. rigidus*，*Intergricorpus bellum*，*Pseudointegricorpus clarireticulatus*。仅见于马斯特里赫特期的孢粉为：*Aquilapollenites funkhouseri*，*A. reductus*，*Pseudoaquilapollenites striatus*，*Pseudointegricorpus clariretuculatus*，*Marsypiletes cretacea*，*Zlivisporites novomexicanus*，*Rugulatisporites quintus*，*Wodehouseia stanley* 等。特别是，*Aquilapollenites reductus* 和 *Marsypiletes cretacea* 等是灭绝于马斯特里赫特期最晚期的分子。该组合与嘉荫邻区、俄罗斯结雅-布列亚盆地的中查加扬组（Middle Tsagayan Formation）孢粉植物组合一致，时代属于马斯特里赫特晚期（Markevich et al.，2011；Tekleva et al.，2015，2019，2020；孙革等，2014）（图24）。

2.2 植物群的性质

嘉荫晚白垩世中—晚期植物群主要由苔藓类、蕨类、银杏类、松柏类及被子植物组成。其中，苔藓类和蕨类均反映了较潮湿的生长环境，它们通常生长在山地谷底或邻近谷底的河湖低地。似里白（*Gleichenites*）的出现似反映气候相对偏湿热，大量的银杏类也反映了喜温喜湿的环境。在松柏类中，大量的杉科植物，如水杉（*Metase-*

黑龙江嘉荫晚白垩世植物群

图24 三棱鹰粉-条纹假鹰粉组合（组合Ⅴ）（据Markevich et al.，2011；孙革等，2014）

Fig. 24 *Aquilapollenites conatus–Pseudoaquilapollenites striatus* Ass.（Ass.Ⅴ）（after Markevich et al.，2011；Sun et al.，2014）

1. *Pseudointegricorpus clariretuculatus*；2. *Aquilapollenites striatus*；3. *A. amplus*；4. *A. rombicus*；5. *A. rigidus*；6. *Marsypiletes cretacea*；7. *Intergricorpus bellum*；8. *I. bertillonites*；9. *Aquilapollenites conatus*；10. *Wodehouseia aspera*

quoia)、红杉(*Sequoia*)、落羽杉(*Taxodium*)及水松(*Gryptostrobus*)等,在我国现今大都生长在暖温带的环境。被子植物中,悬铃木类(如*Platanus*),以及与之相联系的昆栏树(*Trochodendron*)、南蛇藤(*Celastrinites*)及荚蒾(*Viburnum*)等现今也多生长在暖温带,这些植物大都是全缘型、阔叶乔木,它们的落叶性质似乎反映出它们所处的气候环境距亚热带还有一定距离,并具有季节性的变化。大量水生被子植物,如葛赫叶(*Quereuxia*)、卡波叶(*Cobbania*)及莲(*Nelumbo*)等的出现反映当时的水体丰富,有较充足的供水来源。

从植物群中一些被子植物的特征看,嘉荫晚白垩世被子植物的叶片通常较小,如悬铃木(*Platanus*)的叶最长仅6~8 cm,似昆栏树(*Trochodendroides*)叶长一般小于4~5 cm;植物群总体上似具有以小叶型为主的特点。此外,除个别类群出现叶全缘(如悬铃木等)以及革质的叶(如惚木叶*Araliophyllum*等)指示了更温热些的暖温带气候外,大部分被子植物以质地薄且具齿缘叶为主(如似昆栏树等),似显示气候可能相对偏温和湿润,且很可能具有季节性的相对略凉的特点。从这一植物群的晚期组合(太平林场组合)出现一定量的克拉梭花粉(*Classopollis*)看,该植物群在坎潘期可能出现过季节性的干旱。该植物群松柏类的繁盛反映植物群产出的地势可能为山地及坡地,但海拔可能不高。对嘉荫晚白垩世植物群中含有大量喜热喜湿分子,玛尔凯维奇等曾解释其原因:可能嘉荫地区晚白垩世中—晚期由于距海并非很远,受到一定的温暖湿热的海洋性气候的影响(Markevich et al.,2005,2006)。

总体看,嘉荫晚白垩世中—晚期植物群可能是一个暖温带的植物群,具有暖温带落叶阔叶植物群的性质,并具有季节性变化,类似于现今我国长江流域一带低矮山地暖温带的森林。其早期组合(永安村植物组合,桑顿期)可能更偏湿热,而晚期组合(太平林场植物组合,坎潘期)相对温和。这一植物群可能生长在距海不远的内陆低缓的山间盆地,在一定程度上曾受海洋性温暖湿润气候的影响。植被总体性质上,是一个晚白垩世中国北方暖温带的植物群。从大量水生被子植物(如葛赫叶、卡波叶、莲等)的大量存在看,嘉荫地区晚白垩世似水体丰富,雨量较为充沛,为植物生长提供了良好条件。当然,当时这里植物界的繁盛也为恐龙的繁盛提供了有利条件(孙革等,2014)(图25)。

2.3 植物群的时代

根据大植物及孢粉化石的综合研究,嘉荫晚白垩世植物群的时代为晚白垩世的

图 25　东北亚地区晚白垩世植被及古生态环境重建（据 Herman，2011；孙革等，2014）
Fig. 25　Reconstruction of Late Cretaceous vegetation and ecological environment in North-east Asia（after Herman，2011；Sun et al.，2014）

中—晚期，即桑顿期至坎潘期。其时代判定的主要根据包括，植物化石（包括孢粉）自身的地质时代特征，与同期俄罗斯含海相或海陆交互相植物群（包括孢粉植物群）的对比，参考部分同位素测年及与松辽盆地同期地层对比等。

2.3.1　植物化石（包括孢粉）自身的地质时代特征

嘉荫晚白垩世植物群的大部分成员是晚白垩世常见分子，其中还包括一些时代仅限于晚白垩世的重要分子，如达勒比叶（*Dalembia*）、阿朔叶（*Arthollia*）、似南蛇藤（*Celastrinites*）、卡波叶（*Cobbania*）、惚木叶（*Araliaephyllum*）、太平林场似昆栏树（*Trochodendroides taipinglinchanica*）等被子植物，以及裸子植物准落羽杉（*Parataxodium*）及具毛银杏（*Ginkgo pilifera*）等。被子植物悬铃木（*Platanus*）自晚白垩世开始繁盛，并广布于东北亚及北美等晚白垩世地层。其余大部分化石，如裸子植物 *Ginkgo adiantoides*，*Metasequoia disticha*，*Sequoia* sp.，*Glyptostrobus* sp.，*Cupressinocladus sveshnikovae*，被子植物 *Trochodendroides arctica* 及 *Quereuxia angulata* 等也都是晚白垩世常见的化石，尽管它们可延续到古新世或更新的时代。

嘉荫晚白垩世植物群中的孢粉化石的时代指示特征更加鲜明，其大部分组成分子是晚白垩世常见分子，其中还包括一些时代仅限于晚白垩世的重要孢粉化石（图

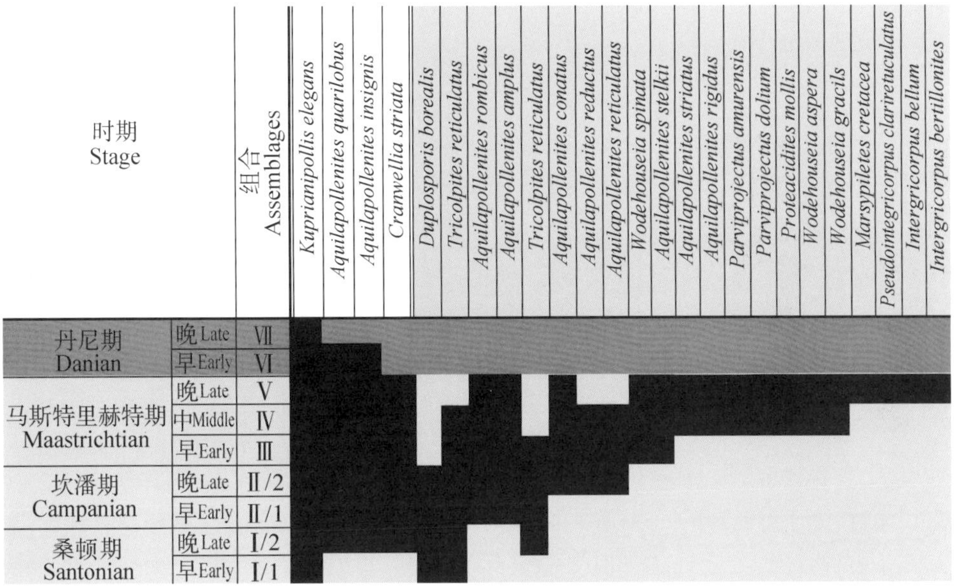

图26　嘉荫及结雅-布列亚盆地晚白垩世重要代表性孢粉地质分布示意图（黑色示地质分布）（据 Markevich et al., 2008）

Fig. 26　Geological ranges of Late Cretaceous important sporopollen in Jiayin and Zeya-Bureya basin（brack color showing the geological distributions）（after Markevich et al., 2008）

26,图27）。从孢粉化石的地质分布图中可以看出,嘉荫仅出现于晚白垩世的孢粉包括: *Aquilapollenites amplus*, *A. conatus*, *A. reductus*, *A. rigidus*, *A. rombicus*, *A. stelkii.*, *A. striatus*, *Intergricorpus bellum*, *Marsypiletes cretacea*, *Parviprojectus amurensis*, *Proteacidites mollis*, *Pseudointegricorpus clarireticulatus*, *Wodehouseia aspera*, *W. gracilis*等,这些都是仅见于晚白垩世的分子;此外, *Kuprianipollis elegans* 及 *Aquilapollenites insignis* 等虽然可延至古新世,但主要分布也是在晚白垩世。值得提及的是,一些晚白垩世的孢粉仅见于马斯特里赫特期, 如 *Aquilapollenites funkhouseri*, *A. reductus*, *Pseudoaquilapollenites striatus*, *Pseudointegricorpus clariretuculatus*, *Marsypiletes cretacea*, *Zlivisporites novomexicanus*, *Rugulatisporites quintus*, *Wodehouseia stanley* 等;其中, *Aquilapollenites reductus*, *Marsypiletes cretacea* 等是灭绝于马斯特里赫特期最晚期的分子(图27)。

2.3.2　与同期俄罗斯含海相或海陆交互相植物群的对比

　　嘉荫北邻,俄罗斯结雅-布列亚盆地含晚白垩世植物群的地层为扎维金组(Zavitin Formation)上部及其相当层位昆都尔组(Kundur Formation)。昆都尔组上部植物组合与太平林场植物组合十分相似,至少含20个共同的分类群,如 *Ginkgo pilifera*, *Se-*

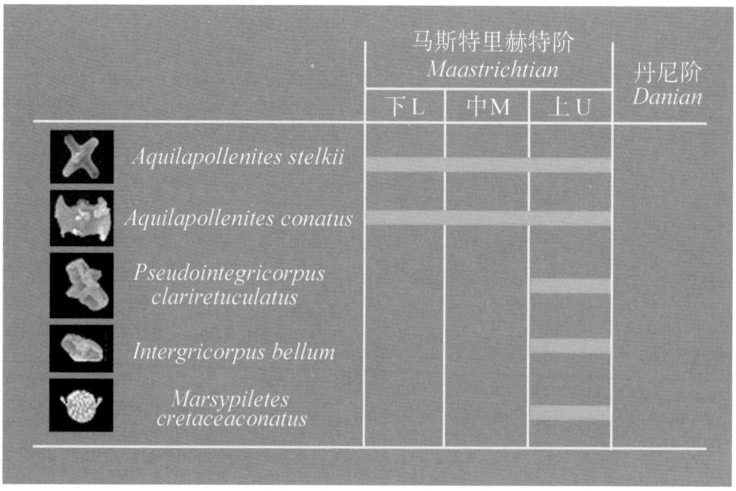

	马斯特里赫特阶 *Maastrichtian*			丹尼阶 *Danian*
	下 L	中 M	上 U	
Aquilapollenites stelkii				
Aquilapollenites conatus				
Pseudointegricorpus clariretuculatus				
Intergricorpus bellum				
Marsypiletes cretaceaconatus				

图27 嘉荫晚白垩世植物群中一些马斯特里赫特期花粉地质分布示意图（据孙革等，2014）

Fig. 27 Geological ranges of some Maastrichtian pollen in Jiayin Late Cretaceous flora（after Sun et al.，2014）

quoia sp.，*Metasequoia* sp.，*Cupressinocladus* sp.，*Trochodendroides lanceolata*，*T. taipinglinchanica*，*Arthollia orientalis*，*Celastrinites kundurensis*，*Quereuxia angulata*，*Cobbania corrugate* 等（Golovneva et al.，2008）；昆都尔的悬铃木类（Platanoids）似更为丰富。其中，*Ginkgo pilifera* 最晚不超过坎潘期（Samylina，1967；Golovneva，2005）；*Celastrinites* 首见于桑顿期，但在马斯特里赫特期最具特征，该属的 *C. septentrionalis*（Krysht.）Gol. 在科里亚克高地（Koryak Upland）繁盛于马斯特里赫特中期的卡诺特组（Kakanaut Formation），该组也产恐龙（Golovneva，1994）。*Cobbania corrugate* 在北美常见于坎潘期—马斯特里赫特期（Bell，1949；Johnson，2002），在西伯利亚北部哈坦加河（Khatanga River）地区也见于坎潘早期（Abramova，1983；Golovneva，2005），那里该层位曾同时发现有坎潘早期的海相瓣鳃类 *Inoceramus patootensiformes*（Golovneva et al.，2008）。考虑到昆都尔组与太平林场组之上均为含恐龙的渔亮子组或下查加扬组，其时代为马斯特里赫特早—中期，因此，太平林场植物组合和昆都尔植物群晚期组合的时代似为坎潘期为宜；孢粉研究结果也认为上述两个组合的时代为坎潘期（Markevich et al.，2005，2006；孙革等，2014）。在东北亚地区，与太平林场组合和昆都尔植物群相似的有库页岛（Sakhalin Island）的宗克尔植物群（Zonk'erian flora，坎潘早期；Krassilov，1979）、科里亚克高地的巴里科夫植物群（Barykovian flora，桑顿期—坎潘早期；Herman & Lebedev，1991），以及穆提诺植物群（Mutino flora，坎潘早期）及西姆植物群（Sym flora，坎潘期，Golovneva，2005）等。

2.3.3 同位素测年及地层对比资料参考

尽管嘉荫晚白垩世植物群目前尚未有直接的同位素测年数据（有关研究目前本书作者正在进行），但相关同位素测年结果以及地层对比，似可以用于对嘉荫晚白垩世植物群及其地层时代确定的参考。其中，最主要的可用于对比的参考资料来自黑龙江省西部的松辽盆地白垩纪地层研究进展（Wan et al.，2013；Xi et al.，2018；席党鹏等，2019）。

据万晓樵等（Wan et al.，2013）和席党鹏等（2019；Xi et al.，2018）结合松辽盆地晚白垩世地层与嘉荫盆地同期地层对比的研究，认为嘉荫永安村组和太平林场组大体上可与松辽盆地的姚家组及嫩江组对比。从本书作者前不久在嘉荫四号靶场的太平林场组新发现的长头松花江鱼（*Sungarichthys longicephalus* Takai）化石看（图5），该化石以往曾发现于吉林前郭伏龙泉组（=嫩江组），这一新发现为嘉荫太平林场组与松辽盆地嫩江组的对比及时代属于坎潘期提供了重要的动物化石证据。嘉荫盆地永安村组和太平林场组的时代大体在桑顿期—坎潘期范围之内，同位素测年值大体在距今8600万~8000万年（86~80 Ma）的范围。上述对比研究结果从一个侧面证明，嘉荫永安村组及太平林场组及其植物组合的时代为桑顿期—坎潘期的结论是较为可靠的。

2.4 植物群最新研究进展

2015~2018年，在国家科技部基础研究专项"中国白垩系—古近系界线研究"、中国地质调查局专项"全国陆相地层划分对比及海相地层阶完善"（白垩系—古近系界线）项目以及国家自然科学基金青年基金等项目共同支持下，本书作者在黑龙江嘉荫晚白垩世中晚期植物群研究中取得重要新进展，主要包括（1）晚白垩世被子植物达勒比叶（*Dalembia*）的发现；（2）晚白垩世水生被子植物莲（*Nelumbo*）的发现；（3）晚白垩世水生被子植物葛赫叶（*Quereuxia*）的叶表皮构造的首次发现等。

2.4.1 达勒比叶在嘉荫的首次发现

达勒比叶是奇数羽状复叶，具5枚小叶。小叶呈椭圆形、卵形或近三角形，顶端钝圆，小叶基部呈楔形、截形或微心形，略不对称；叶缘全缘或裂缺。脉序为羽状或掌状羽状，侧脉直行或半直行。 该属以往仅在俄罗斯远东地区发现，且层位相对较低

（赛诺曼期—康尼亚克期）。达勒比叶在黑龙江嘉荫的发现，首次扩大了对该属时代及古地理分布的认识；该属时代可达到晚白垩世桑顿期，分布范围最南端可达中国黑龙江嘉荫。此成果不仅提高了对该属时代及古地理的研究程度，也为中俄之间东北亚地区晚白垩世植物群及地层的联系与对比提供了新的证据（Sun et al.，2016）（图16，图28）。

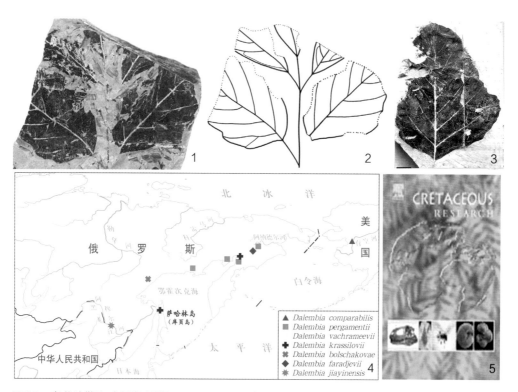

图28　嘉荫达勒比叶新发现（据Sun et al.，2016）

Fig. 28　New discovery of *Dalembia jiayinensis* (after Sun et al.，2016)

1、3.嘉荫达勒比叶化石；2.1的素描图；4.达勒比叶在东北亚地质地理分布图；5.发表杂志封面(2016)

1, 3. Fossils of *Dalembia jiayinensis*; 2. Drawing of 1; 4. Geological and geographic map on distributions of *Dalembia* in NE Asia; 5. Cover of the issue of the paper (2016)

2.4.2　水生被子植物"莲"化石在嘉荫的首次发现

　　2017~2018年，作者在嘉荫晚白垩世永安村组首次发现水生被子植物"莲"化石，并建立嘉荫莲（*Nelumbo jiayinensis* Liang，Sun et Yang）新种。首次发现的嘉荫莲化石保存了较完整的盾圆形叶，叶柄位于叶片中心，全缘叶略显波状，具20~25条一级脉，自叶片中心辐射状向叶边缘伸展，在近叶缘处分叉1~2次；二级脉弯曲并交织成网状;三级脉变弱并织成细网格。上表皮细胞呈较规则的多边形,大小为20~50 μm,垂周壁较直;下表皮细胞与上表皮细胞形态相似,大小为40 μm×30 μm。未见气孔器。成果于2018在 *Cretaceous Research* 发表(图29)。

图29　嘉荫莲及其化石产地

Fig. 29　*Nelumbo jiayinensis* and its locality in Jiayin

1、2.化石产地;3~5.嘉荫莲化石(5为近叶缘的脉序);6.论文发表刊物封面(Liang et al.,2018)

1, 2. Geological and geographic site of *N. jiayinensis*; 3–5. Fossils (5 showing the venation near leaf margin); 6. Cover of the journal of the paper (Liang et al.,2018)

　　莲属植物又称荷花或莲花,是我国现今"十大名花"之一。世界最早的莲化石产自美国及葡萄牙早白垩世阿尔布期,我国以往最早的莲化石记录是发现于海南岛、黑龙江依兰及辽宁抚顺等的始新世地层。因此,嘉荫莲化石的发现将我国莲化石记录至少提前了3000多万年。由于莲属水生植物主要生长在亚热带及暖温带气候下水体资源充沛的环境,本次新发现为恢复黑龙江嘉荫及其中俄边境地区晚白垩世古地理及古气候,深入研究水生被子植物莲属的起源与演化,乃至我国及全球被子植物的起源等,均具有重要意义。

2.4.3　晚白垩世水生被子植物葛赫叶表皮构造首次发现

　　葛赫叶是小型水生被子植物,最早发现于北美白垩纪及古新世地层中。Newber-ry(1861)曾将其当作蕨类,命名为*Neuropteris? angulata*;此后,Lesquereux(1878)及李星学(1959)等将其归于水生被子植物菱属(*Trapa*)。实际上,葛赫叶与菱属植物有明显区别:葛赫叶的浮水叶为莲座状复叶,而菱为单叶;二者的果实也明显不同(Liang et al.,2018)(图30)。基于此,Krystofovich(1953)建议命名这一已绝灭的化石植物为*Quereuxia*。Samylina(1988)首次提出,*Quereuxia*有浮水叶与沉水叶之分(图21,9)。"葛赫叶"这一中文名是孙革根据*Quereuxia*命名人Lesquereux L.(勒士葛赫)的姓氏简化而使用,最早见于全成的博士论文(全成,2006)。此前,张志诚曾将其中文属名称为"奎氏叶"(张志诚,1984)。

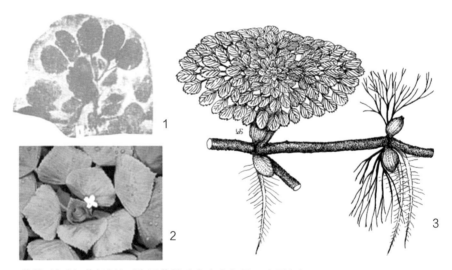

图30　葛赫叶(1)与菱(2)的对比及葛赫叶古生态复原示意图(3)

Fig. 30　Comparison of *Quereuxia* (1) with *Trapa* (2),and ecological reconstruction of *Quereuxia* (3)

　　尽管对葛赫叶化石的研究已有百余年研究历史,但长期以来一直缺乏对其叶表皮构造的研究。本书作者通过运用超景深显微镜观察,首次发现嘉荫晚白垩世的具棱葛赫叶(*Quereuxia angulata*)的浮水叶的表皮构造特征:浮水叶上表皮细胞多呈长方形(50~70 μm×30 μm)或规则的多边形(4~6边),垂周壁略平直,具少量气孔器;气孔器长椭圆形,定向排列,副卫细胞4~6个,保卫细胞外侧可见有纵向褶皱。其下表皮细胞多边形(4~6条边),未见明确的气孔器(梁飞等,2018)(图31)。

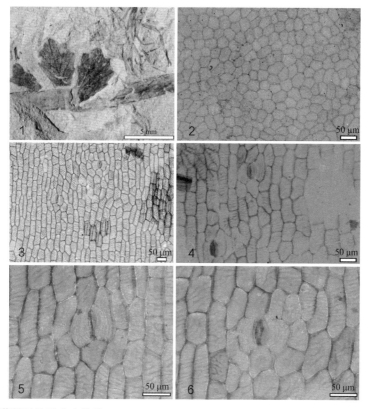

图31　具棱葛赫叶的叶表皮构造

Fig. 31　Cuticles of leaves of *Quereuxia angulata*

1. 叶化石及原位角质层；2. 下表皮细胞；3~6. 上表皮细胞及气孔器（5为3的放大；6为4的放大）

1. Leaves and their cuticles in situ; 2. Lower cuticle; 3–6. Upper cuticles（5 is enlargement of 3; 6 is enlagement of 4）

　　葛赫叶表皮构造的发现对深入研究晚白垩世水生被子植物的分类及当时的古生态环境特征等均具有重要意义。

具棱葛赫叶

Quereuxia angulata

第三章

嘉荫晚白垩世气候与古地理

　　白垩纪(距今约144~66 Ma)是地球构造运动演化最活跃的时期之一,也是地质历史时期生物及其环境特别是古气候演化最重要的时期之一。由于白垩纪全球海平面总体升高,大片陆地被温暖的浅海覆盖,白垩纪全球气温较高,呈现明显的"温室"效应。特别是白垩纪中期,中洋脊的迅速扩张导致海平面上升,使地球变得更加炎热,全球气温可高达现今气温的3~10倍(胡修棉,2004)。当然,白垩纪中期,海洋低层的流动滞缓,造成海洋的缺氧环境,被称为"大洋缺氧事件"(Ocean Anoxic Events, OAE);全球许多黑色页岩层(例如英国北海等)就是在这段时期的缺氧环境中形成的,这些页岩层是当代重要的石油和天然气来源。

　　当然,晚白垩世时期全球的气候也是波动变化的。根据对构造地质、古生物化石及古二氧化碳(pCO_2)等的综合研究表明,晚白垩世早—中期,全球气温一度骤升,出现了"赛诺曼—土伦期极热事件"(Thermal Maximum),大气中的二氧化碳(CO_2)水平也相对较高;但至晚白垩世中—晚期(桑顿期—坎潘期),全球气温有所下降;至白垩纪晚期,气温逐渐降低(Wang et al.,2014;图32)。大量研究表明,白垩纪晚期,随着南大西洋张开、火山活动频繁、大规模海退(Haq,1987)等所造成的陆地生态系统的巨大改变等,陆地上的生物霸主——恐龙等集群灭绝,陆地植物界也发生了巨大的改变,地球上出现了自显生宙以来演化历史中第5次也是最大规模的一次生物灭绝与复苏事件,即白垩纪—古近纪(K-Pg)生物演化事件(孙革等,2014)。

晚白垩纪中—晚期(桑顿期—坎潘期),全球气候总体上相对干燥,但一些濒临海洋地区(如东亚滨太平洋地区,包括俄罗斯远东及我国黑龙江东部等地区)受海洋性气候影响,气温相对有所上升并显得更加湿润,因此,黑龙江嘉荫及其邻区的气候环境更有利于植物界的繁盛与发展。花果繁茂的被子植物的大发展,又为恐龙等陆生动物的空前发展提供了更充足的食物来源。该时期,全球以鸭嘴龙类及角龙类等为代表的植食性恐龙演化与发展达到最高峰;相应的是,肉食性恐龙——暴龙类(如霸王龙等)等,也达到了其发展的极致。在中国,大批恐龙不远千里,集聚到包括嘉荫及俄罗斯结雅-布列亚盆地在内的黑龙江东北部地区,说明这里的气候与古生态环境有利于恐龙等陆生生物繁衍生息。

黑龙江嘉荫地区晚白垩世中—晚期植物群可能正是在这样一个全球性气候及古地理演变大背景下演化与发展的。

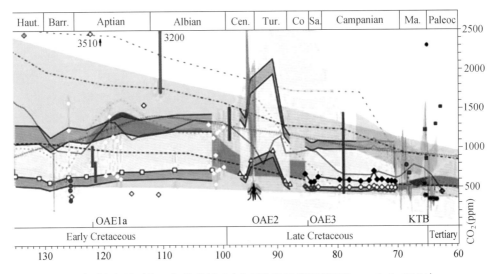

图32 白垩纪全球气候波动与二氧化碳(CO₂)含量变化示意图(据Wang et al.,2014)
Fig. 32 Sketch map showing the changes of Cretaceous climates with CO₂ in the world (after Wang et al.,2014)

3.1 嘉荫晚白垩世气候

3.1.1 植物化石的反映

如同本书第二章所述,嘉荫晚白垩世中—晚期植物群迄今已发现34属43种植物。其中,苔藓类约占2%,蕨类约占11%,银杏类约占7%,松柏类约占27%,被子植物约占50%,被子植物已成为在植物群中占主导地位的优势类群。这里的被子植物、松柏类、银杏类和蕨类等,代表了晚白垩世中期(桑顿期)和晚期(坎潘期)生长在嘉荫地区植被的总体面貌。

从苔藓类和蕨类(包括有节类)来看,它们多生活于空气潮湿的森林中或地下水充足的湿地或沼泽地带,通常反映气候温暖潮湿,但在嘉荫晚白垩世中—晚期植物群的早期组合(永安村组合)中出现以似里白(*Gleichenites*)等为代表的真蕨类,似乎反映至少在晚白垩世中期(桑顿期,即永安村组沉积时期)的气候可能一度相对较炎热。

从银杏类植物看,嘉荫晚白垩世中—晚期植物群中的银杏类化石虽种级分类群不多(仅3种),但数量较大,说明当时在这里银杏类生长较为繁盛。联系现生银杏(*Ginkgo biloba* L.)的生态环境看,现今银杏自然群落多分布于北纬30°附近的暖温带环境;现今银杏类生长最繁茂的地区,如长江流域的洼地或深山山麓,气候温湿,年平均气温9~18 ℃,年平均降水量600~1500 mm(He et al.,1997)。银杏分布的其他地区大体也在我国东部的沈阳以南至福建以北,气候总体上偏温暖潮湿。因此,似可推测,嘉荫地区晚白垩世中—晚期的气候曾是温暖且湿润的。

从松柏类反映的古气候看,现生杉科植物,包括水松(*Glyptostrobus*)、水杉(*Metasequoia*)、落羽杉(*Taxodium*)及红杉(*Sequoia*)等,通常生长在暖温带气候条件下。现存水杉(*Metasequoia glyptostrobus*)目前其自然群落中仅有约5000棵成年植株,主要分布在我国暖温带气候、降水量相对较高的长江流域。红杉为常绿植物,也生长于稳定的温暖潮湿性气候。从嘉荫晚白垩世植物群出现大量杉科植物并且其聚集分布看,似反映晚白垩世中—晚期嘉荫地区的温度和湿度均较现代要高,具有暖温带或暖温带—温带混交的气候。

从被子植物化石看,嘉荫晚白垩世中—晚期植物群中含有大量的悬铃木类(如*Platanus*)、惚木叶(*Araliaphyllum*)、似昆栏树(*Trochodendroides*)、似南蛇藤(*Celastrinites*)、达勒比叶(*Dalembia*)等高大乔木类阔叶落叶被子植物。现生悬铃木类植物多生长于气候温暖且湿润的环境,现今在我国大体分布在北京—大连一线以南、浙闽交界一带以北。嘉荫晚白垩世植物群中水生被子植物的繁盛,如出现大量葛赫叶(*Que-reuxia*)、卡波叶(*Cobbania*)乃至莲(*Nelumbo*)等化石,除反映了这一地区曾水体丰富外,也反映了当时这一地区气候温暖,正如现今的莲属植物(荷花)仅生长在温暖潮湿的气候环境中。

当然,从嘉荫晚白垩世植物群被子植物叶片通常偏小、革质叶不多并多为落叶植物以及植物群含有一定数量的松科(如 *Larix* 等)等状况分析,这一植物群反映了该地区具有季节性的变化(季节性变凉)的特点,其气候总体上可能以暖温带为主,兼有一定的温带气候。

从孢粉化石反映的古气候特征看,嘉荫晚白垩世中—晚期植物群中含有大量喜热、喜湿分子(Markevich et al.,2005),似可以进一步说明嘉荫晚白垩世中—晚期的古气候是以暖温带气候为主。当然,从嘉荫晚白垩世植物群晚期组合(太平林场组合)发现一定数量的买麻藤类花粉化石看,嘉荫地区晚白垩世晚期之初的坎潘期气候可能曾一度略干旱(或季节性的干旱),较晚白垩世中期(桑顿期)气候可能更偏温和。

总之,从在已发现的大量植物化石反映的证据看,晚白垩世的中—晚期,植物群所处的古气候可能为以暖温带为主,兼具季节性变化;植被性质可能类似于现今我国长江流域一带低矮山地的暖温带森林。

3.1.2　古气候定量研究

古植物化石可为研究古气候提供有益帮助。在运用古植物化石叶相分析研究古气候方面,近年来主要有"叶缘分析法"(Leaf Margin Analysis,LMA)、"气候多变量程序分析"(Climate-Leaf Multivariate Program,CLMP)(Wolf,1995;Wolf & Spicer,1999)以及"共存分析法"(Co-existence Approach,CA)(Mosbrugger & Utescher,1997)等。共存分析法是基于"最近现生亲近法"(Nearest Living Relative Method,NLR)研究古气候(Mosbrugger,1999)。近年来,通过植物叶化石气孔器研究古大气二氧化碳浓度进而定量研究古气候,在我国已取得重要进展。这是基于在研究全球碳循环变化(包括地史时期的)对地球生态系统影响中发现,大气中的二氧化碳含量越高,通常反映陆地上的温度越高。化石植物角质层通常保存了叶表皮构造,包括气孔器等。气孔是控制植物与外界进行气体交换(包括交换光合作用中的二氧化碳)的结构,因此,通过计算气孔的一些常数,如气孔指数(SI:stomatal indices)及气孔密度(SD:stomatal density)等,通常可以计算出二氧化碳的浓度。研究表明,地史时期的植物气孔密度及气孔指数均与二氧化碳浓度成反比。由此,通过化石角质层来研究地史时期二氧化碳浓度的变化已成为最有效的方法之一,特别是银杏属化石被认为是用于重建古环境的主要分类群,其研究结果的可靠性更高。

　　近年来,全成等根据采自嘉荫永安村组和太平林场组的大量银杏化石,开展了以高分辨率分析其古二氧化碳(pCO_2)水平的定量研究,报道了对采自嘉荫太平林场组11个层位(每层约5枚叶片)共77枚似铁线蕨型银杏(*Ginkgo adiantoides*)叶化石样品的古二氧化碳浓度系统研究(Quan et al., 2009)(表3)。根据叶角质层气孔指数序列分析的坎潘期大气中的古二氧化碳浓度结果表明,其浓度在0.055%~0.059%(550~590 ppm)范围内逐渐有所降低,这一结果更接近于GEOCARB-Ⅱ模型分析的结果,尽管新数据较之后者的平均值稍高约0.003%(30 ppm)。分析中发现,在坎潘晚期曾出现一个明显的二氧化碳浓度的短期波动(SCDF),古二氧化碳浓度曾达到0.069%(690 ppm),但此后又回到0.059%(590 ppm)的水平。上述研究运用了产自嘉荫地区的单一化石物种——似铁线蕨型银杏。此研究为定量恢复晚白垩世中—晚期古大气中二氧化碳浓度做了有益的尝试。上述研究结果表明,嘉荫地区桑顿期—坎潘期古大气二氧化碳浓度为现代大气二氧化碳浓度的1.4~2.5倍(图33)。相应地,嘉荫

表3　嘉荫晚白垩世太平林场组铁线蕨型银杏气孔器测量及相关二氧化碳浓度(据Quan et al.,2009,简化)

Table 3　Stomatal measurement and inferred paleo-CO_2 based on *Ginkgo. adiantoides* from Cretaceous Taipinglinchang Fm. of Jiayin (after Quan et al.,2009,simplified)

层号 Bed	叶片数 Leaf no.	气孔指数 (平均) SI(mean)	标准误差 Standard deviation of SI	古二氧化碳浓度 pCO_2(RF)	与现今二氧化碳 浓度比率 RCO2
11	10	7.01	0.14	558.53	1.62
10	12	6.98	0.27	569.43	1.62
9	7	6.97	0.43	571.03	1.63
8	8	6.92	0.34	588.02	1.64
7	11	6.91	0.27	592.02	1.64
6	8	6.70	0.23	692.02	1.69
5	3	6.96	0.49	575.98	1.63
4	5	6.98	0.31	567.38	1.62
3	7	6.94	0.31	683.55	1.63
2	5	6.93	0.50	584.90	1.63
1	1	6.83	0.49	624.70	1.66

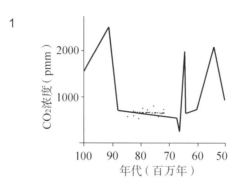

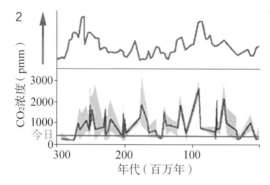

图33 运用银杏化石研究嘉荫桑顿期—坎潘期二氧化碳(CO_2)浓度变化及与北美相关研究对比
Fig. 33 Study of changes of CO_2 concentrations in Santonian–Campanian in Jiayin with fossil *Ginkgo*,and compared to the related study in North America

1. 中国嘉荫研究结果（据Quan et al.,2009）；2. 北美相关研究（据Retallack,2001）

1. Studying result in Jiayin,China（after Quan et al.,2009）; 2. Comparable study result in USA（after Retallack,2001）

晚白垩世中—晚期的气温也远高于现代的气温。这一结果为前述定性分析嘉荫及其邻区晚白垩世中—晚期的古气候,是一个有力的配合。

3.2 嘉荫晚白垩世地理环境

嘉荫晚白垩世植物群及其产出地层等的研究为探讨晚白垩世中—晚期嘉荫地区的古地理环境提供了重要参考资料。

从沉积相反映看,嘉荫及其邻区晚白垩世中—晚期在古沉积环境上曾经历了从河流–湖泊相(以永安村组为代表,图34)—湖泊相(以太平林场组为代表)—河流相(以渔亮子组为代表)—湖泊相(以富饶组为代表)的演变过程。永安村组砂砾岩层的多次出现,反映了这一时期(桑顿期)河流相沉积发育。当然,大量泥岩和粉砂岩的存在也反映湖泊环境可能已初具规模。此期,嘉荫及其邻区的河湖周围地势有一定起伏,很可能为低缓的山地;其坡地上可能生长着繁茂的松柏类(如 *Cupressinocladus*、*Metasequoia*、*Sequoia* 等)、银杏类(如 *Ginkgo pilifera*、*G. adiantoides* 等)以及悬铃木类(Platanoid)被子植物等(孙革等,2014)。2002 年,孙革率领的国际科研队在永安村组中粗粒砂岩中发现恐龙足迹化石——嘉荫足印(董枝明等,2003),进一步提供了永安村组沉积可能反映河流相的证据,因为恐龙等大型爬行类动物通常生活在距森林不远的河湖岸边。以太平林场组为代表的细碎屑沉积(泥岩、页岩及粉砂岩等)反映了这一时期(坎潘期)的湖相分布相当广泛。叶肢介(如俞氏链叶肢介)、介形类(如优雅女星介及年轻刺星介等)以及鱼(如长头松花江鱼)等在太平林场组的发现为当时的湖相环境进一步提供了化石依据。由此推断,嘉荫地区晚白垩世中—晚期可能曾处于较为温湿的环境,气候可能为暖温至中温,降水量可能较高,古地理环境可能包括有浅湖(水体内)、湖岸、沙洲、林前湿地及山前坡地等,其植被随不同的地域形成不同的古植物群落(孙革等,2014)。

　　从古植物化石反映的古地理环境看,晚白垩世永安村组及太平林场组植物组合均显示,当时的植物群可能生活在以暖温带气候为主、兼具季节性变化的气候条件下,处于低矮山地及其环绕的河湖环境。两个植物组合中一些高大的阔叶落叶被子植物(如悬铃木、似昆栏树、似南蛇藤、达勒比叶等)以及大量暖温带松柏类(水杉、红

图34　永安村组的砂岩透镜体及恐龙化石产出反映河流相古环境

Fig. 34　Sandy lens in Yong'ancun Formation and dinosaur track implying a fluvial facies

1. 砂岩透镜体反映河流相沉积;2. 嘉荫恐龙足印的发现(2002)(据孙革等,2014)

1. Sandy lens implying the fluvial facies; 2. Discovery of dinosaur track in Yong'ancun Formation (2002)(after Sun et al.,2014)

杉、水松等)的出现,似反映山地、坡地的古地理环境;葛赫叶、卡波叶及嘉荫莲等水生被子植物化石的大量出现表明,晚白垩世中—晚期的嘉荫地区是水体来源充沛、河湖密布的温暖湿润环境。嘉荫太平林场植物组合时期(坎潘期)嘉荫可能位置更偏于内陆并地势略高,但嘉荫这一时期的山间盆地由于距海不远,得到了相对温暖又湿润的条件。以鸭嘴龙为代表的大量植食类恐龙的伴生也能说明,这里的古地理及古气候是温暖潮湿、水源供应充足的(图35)。

此外,晚白垩世银杏类化石(如 *Ginkgo adiantoides*、*G. pilifera* 等)的大量出现也具有较重要的古生态指示意义。联系现生银杏(*Ginkgo biloba* L.)看,现今银杏自然群落多分布于北纬30°附近的暖温带地区,如长江流域的洼地或深山山麓,气候温暖潮湿,年平均气温9~18 ℃,年平均降水量600~1500 mm(He et al.,1997)。而晚白垩世时期,嘉荫的地理位置已处于北纬40°以北地区,在此能出现大量银杏类(如 *Ginkgo adiantoides* 等),一是说明当时全球气候更温暖,地表温度远高于现代;二是嘉荫及黑龙江东部地区当时可能受到东邻古太平洋暖流的影响,当地的温度和湿度都较现今要高得多。

然而,从嘉荫晚白垩世植物群组成特征的另一方面而言,该植物群中的被子植物叶片通常较小,叶通常非全缘、以质地较薄且具齿缘叶为主。这一特点似表明,尽管

图35 晚白垩世嘉荫地区古地理复原(据孙革等,2014)
Fig.35 Reconstruction of paleogeographic circumstance of Jiayin during Late Cretaceous (after Sun et al.,2014)

嘉荫晚白垩世中—晚期的气候可能以暖温带气候为主,但相对也掺有偏温和的情况,且可能具有一定的季节性变凉的特点。松柏类的大量出现反映当时这里的地势可能相对略高,植物群可能主要生长在坡地及邻近谷底的低地(孙革等,2014)。

从动物化石对古地理环境的指示看,嘉荫及其邻区晚白垩世中—晚期叶肢介、介形类、瓣鳃类及鱼等水生生物繁盛。俄罗斯学者Riabinin(1930)在发表有关满洲龙的论文时,还报道了与恐龙伴生的龟(*Aspederetes planicostatus*)化石的发现,反映了当时湖沼等水体的广泛存在。就恐龙而言,嘉荫及其邻区晚白垩世95%以上的恐龙为植食性的鸭嘴龙类(Hadrosauridae),它们以大量的植物为食,温暖潮湿的环境是这些植物赖以生存和繁盛的保证。嘉荫地区晚白垩世中—晚期,以鸭嘴龙类为主的恐龙在此地区繁衍生息并达到顶峰;这些庞然大物赖以生存的基础——植物界,其繁茂程度,可想而知。大量研究表明,恐龙主要生活在气候温暖潮湿、距森林不远的河湖边缘开阔地带;嘉荫地区大量动、植物化石的发现,为重现嘉荫地区晚白垩世中—晚期古地理环境提供了丰富的参考依据(图35)。

迪克逊铁角蕨
Asplenium dicksonianum

第四章

嘉荫晚白垩世植物与恐龙

恐龙主要生活在湖岸平原的森林地区或植物繁茂的开阔地带。晚白垩世中—晚期,由于黑龙江嘉荫一带植被繁茂、气候温暖湿润,众多恐龙特别是大批植食性恐龙,从遥远的中国西北、华北和东北南部等地区向黑龙江嘉荫迁徙。嘉荫恐龙中,以鸭嘴龙类为代表的植食性恐龙约占95%以上,已发现至少10个分类群(董枝明等,2003;吴文昊等,2010;Godefroit et al.,2011);此外还有少量以暴龙类为代表的肉食性恐龙等(Bolotsky,2013)。鸭嘴龙类恐龙主要包括(图36):

鸭嘴龙亚科 Hadrosaurinae

① 黑龙江满洲龙 *Mandschurosaurus amurensis*(Riab.)Riabinin,1930

② 巨型满洲龙 *Mandschurosaurus magnus*,Zhao,1995(MS)

③ 姜氏嘉荫龙足印 *Jiayinosauropis johnsoni* Dong et al.,2003

④ 马氏克伯龙 *Kerberosaurus manakini* Bolotsky et Godefroit,2004

⑤ 董氏乌拉嘎龙 *Wulagasaurus dongi* Godefroit et al.,2008

⑥ 昆都尔龙 *Kundurasaurus* Bolotsky et Godefroit,2014

图36　黑龙江嘉荫及其邻区晚白垩世的鸭嘴龙类恐龙

Fig. 36　Hardrosaurid of Late Cretaceous dinosaurs in Jiayin and its adjacent area

1. 黑龙江满洲龙；2. 巨型满洲龙；3. 昆都尔龙；4. 阿穆尔龙；5. 姜氏足印；6. 黑龙；7. 大天鹅龙；8. 嘉荫卡龙；9. 乌拉嘎龙；10.克伯龙

1. *Mandschurosaurus amurensis*; 2. *M. magnus*; 3. *Kundurosaurus*; 4. *Amurosaurus*; 5. *Jiayinosauropis johnsoni*; 6. *Sahaliyania*; 7. *Olorotitan*; 8. *Charanosaurus jiayinensis*; 9. *Wulagasaurus*; 10. *Kerberosaurus*

赖氏龙亚科 Lambeosaurinae

⑦ 嘉荫卡龙 *Charonosaurus jiayinensis* Godefroit et al., 2000

⑧ 鄂伦春黑龙 *Sahaliyania elunchunorum* Godefroit et al., 2008

⑨ 里氏阿穆尔龙 *Amurosaurus riabinini* Bolotsky et Kurzanov, 1991

⑩ 阿哈拉大天鹅龙 *Olorotitan arharensis* Godefroit et al., 2003

植食性恐龙包括所有鸟臀类和大部分蜥臀类恐龙,它们通常身体庞大,如我国河南晚白垩世的汝阳龙(*Ruyangsaurus*)体长达38 m,侏罗纪新疆巨龙(*Xinjiangtitan*)体长接近30 m(董枝明,2009;吴文昊等,2013)。如此庞大身躯的恐龙必须有足够的食物供给。据分析,梁龙(*Diplodocus*)的体重约30吨,是现今大象的10倍以上,每日需

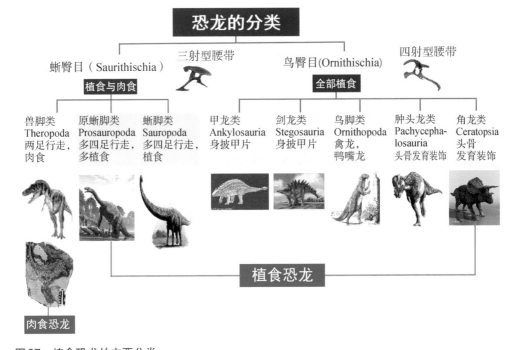

图37　植食恐龙的主要分类

Fig. 37　Main taxonomy of dinosaurs

吃掉植物1~1.5吨。如此巨大的食物量要求植物界必须十分繁茂,才能满足它们的食物供应和生存需求(图37,孙革,2019*)。那么,究竟当时哪些植物曾上过嘉荫植食恐龙的"餐桌"呢?

4.1　植食恐龙的"餐桌"

从形态功能特征看,嘉荫晚白垩世植食性恐龙通常形体庞大,颈部较长,易于向高处取食,如满洲龙、卡龙等身长可达11 m,颈部至少长4~5 m;大天鹅龙形体大,身长约8 m,颈部由18根颈骨组成。这反映当时它们的取食高度有可能达到6~10 m,由此推测它们所吃的部分植物高度至少在6~10 m或更高。植食性恐龙的牙齿多为勺形齿或钉状齿,便于剪断茎叶或刺穿球果外壳(图38,7),说明当时的植被已有具球果类植物,如松柏类及苏铁类等(图38,2、4、8、10)。特别是,鸭嘴龙类恐龙的上、下颌牙齿呈多列排列(如大天鹅龙具齿32列),牙齿数百颗,这些牙齿便于研磨果实(图38,

---

*孙革.中生代植物.沈阳师范大学研究生教材.2019.

图38　嘉荫及其邻区晚白垩世恐龙食用植物分析示意图

Fig. 38　Sketch illustration of analysis on the plants eaten by Cretaceous herbaceous dinosours in Jiayin and its adhacent area

1. 阿穆尔龙与森林（复原图）；2. 巨龙类与松柏类高度示意；3. 鸭嘴龙类与有节类示意图；4. 鹦鹉嘴龙食用苏铁类示意图；5、6. 鸭嘴龙类头骨与牙齿；7. 植食的巨龙类牙齿；8. 现生柏类叶子与球果；9. 现生银杏叶子与种子；10. 现生苏铁类大孢子叶；11. 现生蕨类；12. 现生杉科叶子与果实；13. 鸭嘴龙类与水生被子植物（莲）复原图（据孙革，2019*，有修改）

1. Reproduced paleoenvironment of *Amurosaurus*; 2. Titanosaurid with conifers; 3. Hadrosaurus with horsetails; 4. *Psittanosaurus* eating cycads; 5, 6. Skull and teeth of Hadrosaurid; 7. Teeth of Titanosaurid; 8. Living Cypress; 9. *Ginko biloba*; 10. Megasporophylls of *Cycas*; 11. Ferns; 12. Living *Taxodium*; 13. Reconstruction of dinosaur with aquatic rotus（after Sun, 2019*, with revision）

5、6）。当然，一些植物种皮（如银杏类等）、大孢子叶（如苏铁类等）、大量细嫩的蕨类及被子植物的叶子本身质地就较为柔软，更易于恐龙的咀嚼及消化（图38，3、9、11、13）。

　　从嘉荫地区晚白垩世大量植物化石的发现，以及植食性恐龙的多样性和形态功能等综合分析推测，繁盛的蕨类、多汁的银杏类果子及嫩叶、丰富的松柏类（特别是杉科植物）的果实和叶子，以及芬芳的被子植物的花果和叶子等，都可能曾是嘉荫地区晚白垩世恐龙"餐桌"上的"佳肴"（孙革，2019*）。迪切尔认为，蕨类植物在中生代出现的逐渐矮化和分布范围扩大，可能与恐龙的大量食用有关（Dilcher，2019**）。

———————————————

　　* 孙革. 中生代植物. 沈阳师范大学研究生教材. 2019.

　　** Dilcher. Herbivorous dinosaurs and the plants they ate. Lecture in College of Paleontology, Shenyang Normal University in Shenyang, 2019.8.16.

4.2　植物化石在研究嘉荫K–Pg界线中的作用

白垩系—古近系界线(Cretaceous–Paleogene boundary,简称K–Pg界线,也称K/T界线,K源于希腊文Kreta,即白垩系)是约6600万年(66 Ma)前在地层中留下的一条界线。该界线形成时期,地球上以恐龙为代表的陆生生物及以菊石为代表的海生生物等共约70%的生物发生了大规模的集群灭绝,被称为"显生宙以来地球上发生的第5次生命大灭绝与复苏事件"。就黑龙江嘉荫地区古植物研究而言,孢粉化石保留了充分的证据,从一个侧面揭示了约6600万年前在黑龙江嘉荫及其邻区所发生的这一重大地质事件,特别是这里的生物群在K–Pg界线前后的巨大改变(孙革等,2014)。

K–Pg界线最早由美国科学家阿尔瓦雷斯(L. W. Alvarez)等于1980年首次提出(Alvarez et al.,1980),主要根据他们在欧洲海相白垩系第三系(K/T)界线的黏土层地层中首次发现大量铱(Ir)元素异常。此后,在墨西哥尤卡坦(Yucata)半岛的希克苏鲁伯(Chicxulub)附近的K–Pg界线地层中又发现冲击石英等星际物质。阿尔瓦雷斯等进而提出,K–Pg界线地层中的铱异常和冲击石英等物质来自地外星体对地球的撞击。由此引申,距今约6600万年地球上曾发生"小行星撞击事件",此"撞击"造成了恐龙等生物灭绝。这便是流传甚广的有关恐龙灭绝的"小行星撞击说"。当然,也有许多科学家对此提出不同的意见(Nichols & Johnson,2008),特别是列举了印度德干高原(Deccan Plateau)等在距今约6600万年前曾发生剧烈的火山活动,恐龙灭绝可能与火山的剧烈喷发活动有关(McLean,1985;Keller et al.,2011)。由此,有关K–Pg界线的研究成为全球中生代与新生代之交生物演化与环境变化研究的一个"热点"。

黑龙江嘉荫地区陆相白垩纪—古近纪地层及生物群十分发育,大量的恐龙、鱼、叶肢介、介形类以及植物等化石,记录了黑龙江嘉荫及其邻区晚白垩世中—晚期(距今约8500万~6600万年)丰富多彩的生物世界,也记录了主要由植物等化石提供的协同演化及其古生态环境的变化。沿黑龙江右岸晚白垩世—古新世地层出露较好,化石也十分丰富,为在这里研究K–Pg界线提供了得天独厚的条件。2002~2018年,我国学者孙革率领的国际科研队在嘉荫地区首次发现:(1)以鸭嘴龙类为主要代表的晚白垩世恐龙动物群最晚仅在马斯特里赫特早—中期出现,但马斯特里赫特晚期(即富饶组时期)并未发现恐龙化石,这似乎表明马斯特里赫特晚期(距今约6800万~6600万年),嘉荫及其邻区的恐龙已经灭绝(孙革等,2014);(2)在嘉荫乌云小河沿实施的多个勘探钻孔的地球化学研究表明,6600万年前,在嘉荫及其邻区发生的K–Pg界线的地层中没有见到有意义的"铱异常",也没有见到任何"星际物质",由此推测,这里在K-Pg界线时期并未发生过"小行星撞击事件"。上述成果为研究我国东北乃至整个东北亚地区晚白垩世生物群演化和地球环境变化等提供了宝贵的证据和启示(孙革等,2014)。

4.2.1　嘉荫K-Pg界线的确定

　　2002~2011年,孙革等通过运用古生物学、地层学、同位素年代学、古地磁学以及地球化学等综合研究的方法,特别是依据高分辨率的孢粉地层精细划分,确定将黑龙江嘉荫小河沿XHY-2006钻孔点(129°35′13.5″E,49°14′53.9″N)作为陆相K-Pg界线地层的标准点位,其K-Pg界线确定于该钻孔的22.00~22.05 m之间(Sun et al.,2011;孙革等,2014)(图39)。

　　嘉荫小河沿K-Pg界线的确定,主要是依据嘉荫小河沿XHY-2005、XHY-2006及XHY-2008等三个钻孔完整的钻孔剖面所发现的丰富的孢粉化石证据:K-Pg界线之

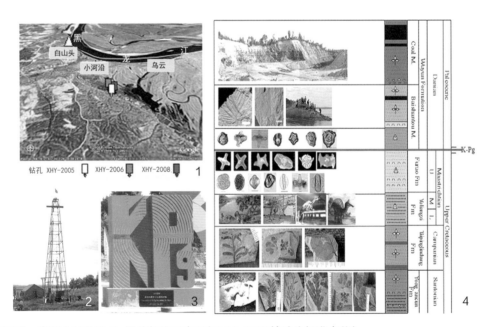

图39　嘉荫小河沿K-Pg界线标准示意图(XHY-2006钻孔为标准点位)

Fig. 39　Illustration of the stratotype of the KPgB in Xiaoheyan of Jiayin (Borehole XHY-2006 as the stratotype point)

1~3.钻孔及点位示意;4.K-Pg界线综合柱状示意图

1-3. The borehole XHY-2006 and its geographic illustration; 4. Sketch stratigraphic column of the K-Pg boundary strata

下为晚白垩世马斯特里赫特期最晚期的孢粉组合"三棱鹰粉-条纹假鹰粉(*Aquilapollenites conatus-Pseudoaquilapollenites striatus*)组合"(组合Ⅴ，产于富饶组)，而K-Pg界线之上为古新世丹宁早期的孢粉组合"混杂三孔庭粉-小刺鹰粉(*Triatriopollenites confusus-Aquilapollenites spinulosus*)组合"(组合Ⅵ，产于乌云组白山头段)。这一界线上下所显示的孢粉组合特征，与嘉荫北邻约30 km的俄罗斯结雅-布列亚盆地白山出露的K-Pg界线及其相关地层等的孢粉组合特征几乎完全一致。白山K-Pg界线剖面是国际公认的、全球重要候选标准剖面之一(Nichols & Johnson，2008)，该K-Pg界线附近地层的孢粉化石经玛尔凯维奇等多年研究，其组合及界线划分的研究已较为详尽(Markevich et al.，2005，2006)。玛尔凯维奇等将白山地区K-Pg界线附近地层的孢粉植物群划分为7个组合，其中组合Ⅴ及Ⅵ是划分K-Pg界线的标志(Markevich et al.，2006；IBP et al.，2001)。上述两个孢粉组合的组成特征分别与嘉荫小河沿三个钻孔岩心所显示的K-Pg界线之上及之下的孢粉组合几乎完全一致(孙革等，2014)。

其他证据还包括同位素测年、古地磁以及地球化学等研究成果。嘉荫K-Pg界线之上的及古新世早期白山头段的酸性凝灰岩U-Pb法测年结果为64.1±0.7 Ma(Suzuki et al.，2011)，确属古新世丹尼期(Danian)。相关对比测年也曾在俄罗斯白山剖面同期火山凝灰岩开展，LA-ICP-MS法锆石测年结果为66±1 Ma，结合其上、下地层含孢粉化石的时代，认为属于古新世丹尼早期(Knittel et al.，2013)。古地磁学研究方面，根据对保存完整的XHY-2006及XHY-2008两个钻孔岩心样品的磁倾角变化及正反极测试结果，嘉荫小河沿的K-Pg界线应在C29r(65±0.3 Ma)的反极性区；XHY-2006及XHY-2008两钻孔古地磁学研究确定的K-Pg界线分别位于22.05 m和23.25 m附近，与孙革及玛尔凯维奇等根据孢粉化石确定的界线位置基本吻合(Sun et al.，2011；孙革等，2014)(图40)。

地球化学分析结果表明，嘉荫小河沿K-Pg界线附近地层不存在有意义的"铱异常"。

4.2.2 植物化石在研究K-Pg界线中的作用

尽管嘉荫的植物大化石受保存及界线确定精度等限制，未能在精确确定K-Pg界线中发挥直接作用，但植物微体化石(孢子与花粉化石)在嘉荫的K-Pg界线划分研究中发挥了关键性重要作用。

嘉荫小河沿K-Pg界线附近地层神奇地保存了晚白垩世最晚期至古新世最早期的孢粉化石，玛尔凯维奇等孢粉学家从嘉荫小河沿上述3个钻孔(XHY-2005、XHY-

黑龙江嘉荫晚白垩世植物群

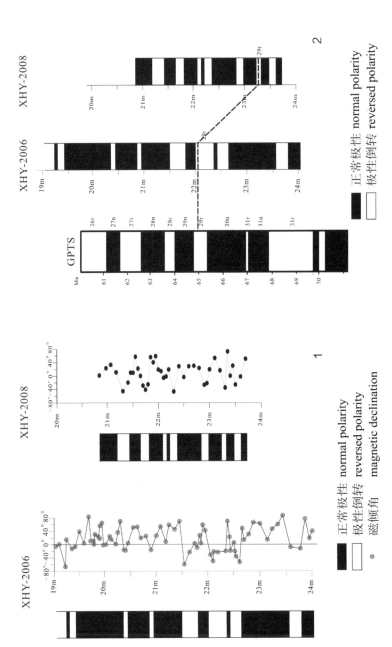

图40 嘉荫小河沿K–Pg界线的古地磁研究

Fig. 40 Paleomagnetic study of the K–Pg boundary in Xiaoheyan of Jiayin

1. 钻孔岩心磁倾角变化及其正反极性状图；2. 极性倒转对比图（据孙革等，2014）

1. Histograms of XHY–2006 and 2008, showing paleomagnetic declination and polarity; 2. Polarity reversal histogram（after Sun et al., 2014）

2006、XHY-2008)共200余件孢粉样品中,以"厘米级"的精度,高分辨率地识别出K–Pg界线的存在:这里的K–Pg界线之下的孢粉组合为典型的晚白垩世马斯特里赫特晚期三棱鹰粉–条纹假鹰粉孢粉组合(组合V),其产出地层为晚白垩世最晚期的富饶组的顶部;K–Pg界线之上为确切的古新世丹尼早期混杂三孔庭粉–小刺鹰粉孢粉组合(组合Ⅵ),产出地层为古新世最早期的乌云组白山头段的底部;K–Pg界线即划分于第Ⅴ孢粉组合地层与第Ⅵ孢粉组合地层之间(Markevich et al.,2011;Sun et al.,2011)(图41)。在XHY-2006钻孔,K–Pg界线位于该孔岩层的22.00~22.05 m;由于岩心剖面完整、化石丰富,此孔的岩心剖面被选定为嘉荫小河沿K–Pg界线标准剖面("正层型剖面");XHY-2005钻孔(界线位于19.90~20.30 m)和XHY-2008钻孔(界线位于23.05~23.25 m)被选定作为嘉荫小河沿K–Pg界线的"副层型剖面"(孙革等,2014)(图42)。

　　详细研究表明,上述K–Pg界线之下的孢粉组合Ⅴ(三棱鹰粉–条纹假鹰粉组合)为典型的晚白垩世马斯特里赫特晚期孢粉组合。该组合的孢子主要为光面单缝孢;花粉以杉科和榆科的繁盛为特征。被子植物花粉中,鹰粉(Aquilapollenites)达14种之多,其中,*Aquilapollenites stelkii*、*A. conatus*、*Pseudointegricorpus clarireticulatus*、*Marsypiletes cretacea*、*Integricorpus bellum*等的时代仅限于马斯特里赫特期或马斯特里赫特晚期。这一孢粉组合与结雅–布列亚盆地、萨哈林群岛、尤里岛等马斯特里赫特晚期孢粉组合相一致,后两个地区的孢粉组合地层均有海相化石佐证。

　　在界线之上的孢粉组合Ⅵ(混杂三孔庭粉–小刺鹰粉组合)中,一些特征花粉*Triatriopollenites confusus*(混杂三孔庭粉)、*T. plectosus*(普氏三孔庭粉)、*Aquilapollenites spinulosus*(小刺鹰粉)及*A. subtilis*(细弱鹰粉)等时代仅限于丹尼早期。上述组合与萨哈林群岛及科里亚克高地丹尼早期的孢粉组合也十分相似,其时代有海相瓣鳃类及放射虫等证明。特别需要指出的是,在XHY-2005钻孔保存的丹尼早期孢子组合中,出现明显的"蕨类异常丰富"(fern spike)的特征,与在结雅–布列亚盆地显示的一致,为K–Pg界线确定进一步提供了依据(Markevich et al.,2011;Sun et al.,2011;孙革等,2014)。

　　尽管植物大化石尚未能在嘉荫K–Pg界线研究中发挥直接作用,但对识别嘉荫地区K–Pg界线上、下生物群及地层的总体面貌及时代等,发挥了一定的辅助作用。例如,本书作者等曾在嘉荫小河沿附近的白山头的乌云组白山头段发现古新世重要植物大化石——查加扬椴叶[*Tiliaephyllun tsagajanicum*(Krysht. et Baik.)Krassilov]等,该化石曾发现于俄罗斯白山剖面K–Pg界线之上的黄灰色砂岩中,旁证了嘉荫乌云组

黑龙江嘉荫晚白垩世植物群

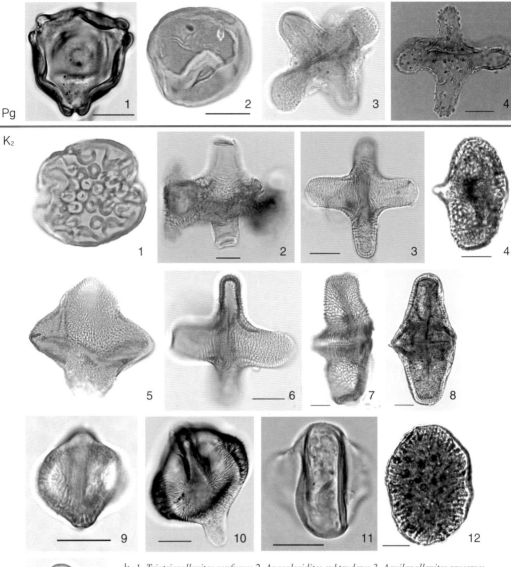

Pg

K₂

上：1. *Triatriopollenites confusus*; 2. *Anacolosidites subtrudens*; 3. *Aquilapollenites proceros*;
4. *Aquilapollenites spinulosus*

下：1. *Marsypiletes cretacea*; 2. *Aquilapollenites conatus*; 3. *Aquilapollenites stelkii*;
4. *Aquilapollenites reticulatus*; 5. *Fibulapollis rombicus*; 6. *Aquilapollenites striatus*;
7. *Intergricorpus bellum*; 8. *Pseudointegricorpus clariretuculatus*; 9. *Aquilapollenites rombicus*;
10. *Triprojectus amoenus*; 11. *Obiculapollis lucidus*; 12. *Wodehouseia aspera*;
13. *Quercoidites minor*

图41　嘉荫小河沿 K-Pg 界线划分的孢粉化石证据

Fig. 41　Palynological evidence for definition on the K-Pg boundary in Xiaoheyan of Jiayin

上：丹尼期孢粉组合（组合Ⅵ，古新世乌云组白山头段）；下：马斯特里赫特最晚期孢粉组合（组
合 V，晚白垩世富饶组）（据 Markevich et al., 2011；Sun et al., 2011）

Up: Danian palynological assemblage（Ass. Ⅵ; Baishantou Mem. of Wuyun Fm.）; Low: Late Maastrichtian palynological assemblage（Ass. Ⅴ; Late Cretaceous Furao Fm.）（after Markevich et al., 2011;
Sun et al., 2011）

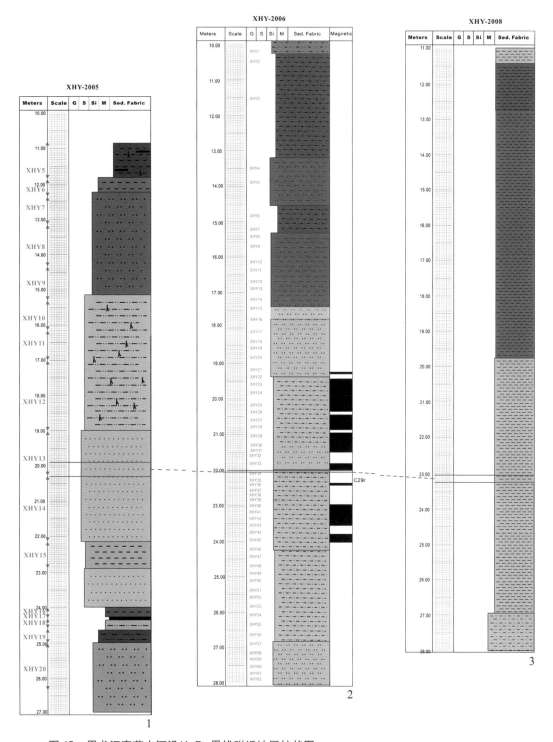

图42　黑龙江嘉荫小河沿K-Pg界线附近地层柱状图

Fig. 42　Profiles of the K-Pg boundary strata in Xiaoheyan of Jiayin

1. XHY-2005孔，界线位于19.90~20.30 m；2. XHY-2006孔（层型剖面），界线位于22.00~22.05 m；

3. XHY-2008孔，界线位于23.05~23.25 m。横向红线示K-Pg 界线（据Sun et al., 2011）

1. XHY-2005, the K-PgB in 19.90-20.30 m；2. XHY-2006, stratotype, the K-PgB in 22.00-22.05 m；

3. XHY-2008, K-PgB in 23.05-23.25 m. Horizontal red lines showing the K-PgB (after Sun et al., 2011)

白山头段属于丹尼期（Sun et al., 2002；孙革等, 2005）；在K-Pg界线之下、整个晚白垩世地层及生物群的研究中，对桑顿期永安村组及坎潘期太平林场组的植物群及地层研究及其晚白垩世中—晚期时代的认定，为研究嘉荫小河沿晚白垩世最晚期的植被变化及K-Pg界线前的生物及其地质背景等，做了较好的铺垫。

第五章

中国东北晚白垩世植物群演化序列

　　由于化石保存及出露条件等原因,迄今我国晚白垩世植物化石点发现较少,在我国东北,以往主要仅在黑龙江省牡丹江(张武等,1980;张志诚,1981)、嘉荫(张志诚,1984;陶君容,2000;孙革等,1995,2014;Sun et al.,2007,2011,2016;全成,2006;Quan et al.,2008,2009;公繁浩,2007;梁飞、孙革,2015;Liang et al.,2018)、松辽盆地(郑少林等,1994)及七台河(Sun et al.,2000,2019)等地点有少量报道。总的看来,晚白垩世植物研究在中国中生代植物研究中是一个"短板",特别是,对晚白垩世植物群的演替序列的研究迄今尚未能如愿完成。

　　然而,晚白垩世各时期植被面貌及其演化序列研究关系到黑龙江地区乃至整个东北地区晚白垩世地层划分与对比,进而或关系到黑龙江省东部地区油气资源的找寻。多年来,我国学者对黑龙江省东部晚白垩世地层划分与对比长期有争议(周志炎、李佩娟,1980;张志诚,1981,1984;黑龙江省地质矿产局,1993;陈丕基,2000;郝诒纯等,2000;孙革、郑少林,2000;孙革等,2014;席党鹏等,2019;Sun et al.,2019),其主要原因还是在于化石找寻特别是晚白垩世植物化石找寻及研究方面存在欠缺。

　　尽管如此,随着近年来我国科学工作者的努力和国际合作研究的加强,目前对我国东北、特别是黑龙江晚白垩世植物群的演化的大体轮廓,已有了初步的认识;特别是在晚白垩世中—晚期植物群(嘉荫)及晚白垩世早期植物群(牡丹江—七台河)研究中取得重要进展(Sun et al., 2007, 2016, 2019; Golovneva et al., 2008; 全成, 2006; Quan et al., 2008, 2009; Liang et al., 2018)。据此,本书作者提出了目前已知中国东北晚白垩世植物演替序列的总体框架(详见5.5)。

需要说明的是,中国东北晚白垩世植物群是作为一个植物群(flora)整体看待,因此,本书以下所列述的各时期"植物群"实际上只是晚白垩世植物群在演化序列上的一个时期的植物组合或"小植物群"(florule)。为称呼上的方便,本书中文仍称各时期植物群为"植物群",但在英文说明中其等级只相当于"小植物群"。

5.1　晚白垩世早期植物群——牡丹江植物群

本书将黑龙江晚白垩世早期植物群统称为"牡丹江植物群"(Mudanjiang florule)。主要化石包括,牡丹江地区晚白垩世植物化石(张志诚,1981),松辽盆地东缘、安达地区泉头组三、四段植物化石(郑少林、张莹,1994),以及七台河七峰林场赛诺曼期植物化石(孙革、郑少林,2000; Sun et al., 2019)等。上述化石显示的地质时代总体上为晚白垩世赛诺曼期(Cenomanian),部分可能延至土伦早期(early Turonian)。这一时期植物群总体组成为:石松类 *Selaginella suniana* Zheng et Zhang;真蕨类 *Cladophlebis* sp., *Onychiopsis psilotoides* (Stocks et Webb) Ward;松柏类 *Thuja heilongjiangensis* Zheng et Zhang;被子植物 *Platanus appendiculata*, *P. heilongjiangensis* Sun et al., *P. pseudiguillelmae*, *P. subnobilis*, *Platanus* sp., *Aralia mudanjiangensis* Zhang,以及 *Dicotylophyllum* sp. 等。

特别值得提及的是,孙革等在七台河七峰林场地区首次发现黑龙江悬铃木等(*Platanus* sp., *P. heilongjiangensis*)化石,并在与化石所在地层整合接触的上部火山岩中测得96.2±1.7 Ma同位素年龄,佐证了该含化石地层时代为赛诺曼期,这为确认牡丹江植物群的时代主要为赛诺曼期提供了重要参考依据(孙革、郑少林,2000; Sun et al., 2019)(图43)。

牡丹江植物群的特征主要为:蕨类植物较少,松柏类发现不多(或许与化石采集有关),但被子植物大量出现特别是悬铃木类植物(如 *Platanus*)繁盛尤为显著。晚白垩世早期植物群迄今仅发现于我国黑龙江省东部及中部,这似乎表明,这一时期我国乃至东亚的大部分地区可能正处于炎热和干旱气候,只是在邻近滨太平洋近海的黑龙江地区生长着一批暖温带针叶—阔叶林植被(张志诚,1981; Sun et al., 2019)。迄

图43　黑龙江晚白垩世早期牡丹江植物群（被子植物部分）

Fig. 43　Early Late Cretaceous Mudanjiang florule of Heilongjiang（angiospermous part）

1、2. 黑龙江悬铃木；3、4、8~11. 悬铃木（未定种）；5. 假奎列尔玛悬铃木；6. 亚显赫悬铃木；7. 灰脉悬铃木；12. 双子叶叶（未定种）；13. 牡丹江惚木

1, 2. *Platanus heilongjiangensis*; 3, 4, 8–11. *Platanus* sp.; 5. *P. pseudiguillelmae*; 6. *P. subnobilis*; 7. *P. appendiculata*; 12. *Dicotylophyllum* sp.; 13. *Aralia mudanjiangensis*

今已知牡丹江植物群所处的地层主要为：黑龙江省猴石沟组上部，以及泉头组三、四段等。

5.2　晚白垩世"中期"植物群

　　本书所指的晚白垩世"中期"植物群，是指晚白垩世偏中期，即土伦期（Turonian）至康尼亚克期（Coniacian）的植物群（florule）。由于晚白垩世中期偏晚的桑顿期（Santonian）的植物群面貌目前已经清楚（以嘉荫永安村组植物组合为代表），而土伦期属于晚白垩世早期，因此，本书所研究的晚白垩世"中期"植物群，实际上所指的是晚于晚白垩世早期（牡丹江植物群）又早于晚白垩世中—晚期（嘉荫植物群）的一个过渡性植物群。

　　黑龙江省境内的土伦晚期—康尼亚克期（late Turonian-Coniacian）植物群的面貌

目前尚不清楚,但这一时期的地层在黑龙江省西部已明确为松辽盆地东缘的青山口组(距今90.0~86.3 Ma)等。这一时期的地层中,目前虽有部分孢粉化石发现(Xi et al.,2018),但可靠的植物大化石还有待找寻。

以往在松辽盆地东缘姚家组发现的化石,如 *Trapa angulata*（Newb.）Brown 等(郑少林等,1994)实为水生被子植物 *Quereuxia angulata*（Newb.）Krysht,似应属晚白垩世中—晚期植物群的范畴;姚家组的时代(距今86.3~84.2 Ma)可能大体与嘉荫永安村组的时代相当(席党鹏等,2019;孙革等,2014)。

5.3 晚白垩世中—晚期植物群——嘉荫植物群

黑龙江晚白垩世中—晚期植物群为嘉荫植物群(Jiayin florule),主要以嘉荫晚白垩世永安村组植物组合及太平林场植物组合的共同组成为代表,即包括:①晚白垩世中期永安村组植物组合[准落羽杉-莲(*Parataxodium-Nelumbo*)组合,时代为桑顿期(Santonian)];②晚白垩世晚期太平林场组植物组合[水杉-似昆栏树-葛赫叶(*Metasequoia-Trochodendroides-Quereuxia*)组合,时代为坎潘期(Campnian)]。

晚白垩世中期永安村组植物组合,迄今已发现24属27种以上,主要包括:

蕨类 *Equisetum* sp.,*Asplenium dicksonianum* Heer,*Arctopteris* sp.,*Cladophlebis* sp.,*Gleichenites* sp.;银杏类 *Ginkgo adiantoides*（Ung.）Heer,*G. pilifera* Samylina;松柏类 *Cupressinocladus sveshnikovae* Ablajev,*Metasequoia disticha*（Heer）Miki,*Sequoia* sp.,*Parataxodium* sp.,*Elatocladus* sp.;被子植物 *Dalembia jiayinensis* Sun et Golovneva,*Menispermites* sp.,*Trochodendroides arctica*（Heer）Berry,*Nyssidium arcticun*（Heer）Iljinskaja.,*Platanus* sp.,*Viburnophyllum* sp.,*Dicotylophyllum* sp.,*Quereuxia angulata*（Newb.）Krysht.,*Cobbania corrugata*（Lesq.）Stockey et al.,*Nelumbo jiayinensis* Liang et al. 等。

晚白垩世晚期太平林场组植物组合,迄今已发现30属38种,包括:苔藓类 *Thallites* sp.;蕨类 *Equisetum* sp.,*Asplenium dicksonianum* Heer,*Cladophlebis* sp.;银杏类 *Ginkgo adiantoides*（Ung.）Heer,*G. pilifera* Samylina;松柏类 *Taxodium olrikii*（Heer）Brown,*Metasequoia disticha*（Heer）Miki,*Sequoia* sp.,*Pityophyllum* sp.,*Pityospermum* sp.,*Glyptostrobus* sp.,*Larix* sp.,*Elatocladus* sp.;被子植物 *Araliaephyllum?* sp.,*Arthollia orientalis*（Zhang）Golovneva,*A. tschernyschewii*（Kostanov）Golovneva,Sun et Bugdaeva,*Celastrinites kundurensis* Gol.,Sun et Bugd.,*Platanus densinervis* Zhang,*P. sinensis* Zhang,*Platanus* sp.,*Trochodendroides arctica*（Heer）Berry,*T. taipinglinchanica* Gol.,Sun et Bugd.,*T.*

lanceolata Gol.、*T. microdentatus*（Newb.）Krysht.、*Viburnum* cf. *contortum* Berry、*Viburno-phyllum* sp.、*Quereuxia angulata*（Newb.）Krysht.、*Cobbania corrugata*（Lesq.）Stockey et al.等。

　　晚白垩世中—晚期（嘉荫）植物群已具有如下特征：①被子植物在组成中的比例已明显增加，在整个植物群中已占优势地位（约占50%以上），尽管仍有悬铃木（*Plata-nus*）、惚木叶（*Araliaephyllum*）及翅子（*Menispermites* sp.）等时代偏早的分子存在，但也出现达勒比叶（*Dalembia*）、阿朔叶（*Arthollia*）、似南蛇藤（*Celastrinites*）等特有的晚白垩世中—晚期被子植物分子（图44）；②裸子植物中首次出现水杉（*Metasequoia*）、红杉（*Sequoia*）等植物，显示出植被组成在逐渐向新生代过渡的特征；③喜温喜湿植物仍占较大比例（如银杏类和蕨类等）；④大量水生被子植物葛赫叶（*Quereuxia*）、卡波叶

图44　黑龙江晚白垩世中—晚期植物群——嘉荫植物群

Fig. 44　Middle−Late Cretaceous Jiayin florule of Heilongjiang

1、2. 蕨类；3. 银杏类；4~6. 松柏类；7~16. 被子植物

1, 2. ferns; 3. ginkgos; 4–6. conifers; 7–16. angiosperms

（*Cobbania*）、莲（*Nelumbo*）等的出现,反映此间水体丰富,为雨量充沛的暖温带—温带的气候环境。

5.4　晚白垩世最晚期植物群——"渔—富"植物群

晚白垩世最晚期(马斯特里赫特期)植物群以嘉荫地区的渔亮子组(马斯特里赫特早—中期)及富饶组(马斯特里赫特晚期)植物组合为代表,本书将其初步命名为"渔—富植物群"(Yu-Fu florule；"渔"指渔亮子组,"富"指富饶组),时代为整个马斯特里赫特期。

这一植物群虽暂时尚无植物大化石的报道,但从丰富的孢粉化石中已经大体得知它们的分类组成及特征(参见2.1.2,③~⑤)。渔—富植物群早期(即渔亮子组植物组合)特征为:蕨类植物仍较发育；裸子植物以松科和杉科占优势,但银杏类及买麻藤类增加(买麻藤类已有6个种)；被子植物主要以榆科、胡桃科(Juglandaceae)、桦科、山毛榉科(Fagaceae)、杜鹃花科(Ericaceae)及杨梅科(Myricaceae)等常见,且多样性增加。渔—富植物群晚期(即富饶组植物组合)特征为:蕨类植物仍较发育；裸子植物以杉科占优势(约占50%)；被子植物主要以榆科、桦科、杨柳科、胡桃科、杨梅科和桃金娘科(Myrtaceae)等为代表,已见桑寄生科(Loranthaceae)、山龙眼科(Proteaceae)及木兰科(Magnoliaceae)分子,但数量较少,被子植物仍然显示了较为丰富的多样性(Markevich et al.,2011；孙革等,2014)。从榆科及桦科等温带植物的出现及含量不断增长看来,渔—富植物群似代表气候适度温和的植物群,较嘉荫中—晚期(桑顿期—坎潘期)植物群所处的古气候至少在温度上已有所降低。

5.5　中国东北晚白垩世植物群演化序列框架

综上所述,尽管迄今中国东北地区晚白垩世植物群的演化序列还处于尚未完善的现状,但东北地区(以黑龙江东部为代表)晚白垩世植物群的演化序列,目前大体可初步总结为:

① 晚白垩世早期植物群——牡丹江植物群。以牡丹江地区晚白垩世植物群为代表,时代以赛诺曼期(Cenomanian)为主,气候似显示高热。

② 晚白垩世"中期"植物群。时代大约以土伦期—康尼亚克期(Turonian-Coniacian)为代表,目前尚在研究中。

③ 晚白垩世中—晚期植物群——嘉荫植物群。以嘉荫地区晚白垩世植物群为代表,时代为桑顿期—坎潘期(Santonian-Campanian),气候似显示暖温带特征。

④ 晚白垩世最晚期植物群——"渔—富"植物群。时代为整个马斯特里赫特期(Maastrichtian),气候似显示已偏温和,具有暖温带—温带的特征。植物大化石有待进一步研究。

目前中国东北地区晚白垩世植物群发展序列的总体框架可参见表4。

表4　中国东北晚白垩世植物群发展序列概示

Table 4　Outline of developmental sequences of Late Cretaceous floras in NE China

时代期次			期名	标准测年(Ma)*	各期植物群	主要特征	备注
晚白垩世	晚	6	马斯特里赫特期	66.0 72.1	渔—富植物群	典型马斯特里赫特期孢粉	早中期恐龙繁殖
		5	坎潘期	84.2	嘉荫植物群	达勒比叶、阿朔叶、似昆栏树、水杉等共同出现,水生被子植物繁盛;被子植物已约占50%	植食恐龙繁盛
	中	4	桑顿期	86.3			
		3	康尼亚克期	89.8	?	?	
	早	2	土伦期	93.9		?	
		1	赛诺曼期	100.5	牡丹江植物群	悬铃木类繁盛	已获年龄参考值96.2 ± 1.7 Ma

*标准测年值据席党鹏等,2019

总之,目前黑龙江晚白垩世植物发展序列研究取得的主要进展为:

1. 黑龙江晚白垩世偏早期(以赛诺曼期为主)植物组合面貌以及晚白垩世中—晚期(桑顿期—坎潘期)植物组合面貌目前已初步探清,其组成内容不断有新的发现,研究也不断深入。

2. 植物群演化中的一个突出特征是,被子植物(特别是水生被子植物)自晚白垩世早至晚期不断发展,自桑顿期起,被子植物在整个植物群中已占主导地位(比例大体约为50%)。

3. 黑龙江东部及东北部地区,伴随植物群演化的古气候演化的总体特征似显示为:晚白垩世偏早期(赛诺曼期)气候偏高热,中晚期气候转为温暖且湿度增加。

有关黑龙江晚白垩世植物演化序列研究,目前存在的问题和未来工作重点似主要为:

1. 晚白垩世偏中期的植物群(大体以土伦期—康尼亚克期为代表)的面貌及特征亟待深入研究。

2. 相关地层(土伦阶—康尼亚克阶)的研究工作也亟须同步进行,找寻相关火山岩、深入开展同位素测年研究等是重要途径。

第六章

嘉荫古植物研究相关科普活动

在国家及黑龙江省各级政府以及社会各界大力支持下,近20年来,嘉荫以恐龙、古植物及K-Pg界线为代表的地质古生物学研究成果不断涌现,"中国第一龙乡""中华第一莲""中国首条陆相K-Pg界线"等美誉给黑龙江嘉荫的发展带来新的生机;俄、德、美、英、法、比、日等多国科学家及学生的到来使嘉荫的地质古生物科普活动不断掀起新的热潮。

嘉荫的地质古生物科普工作有三个"亮点"尤其值得称赞:一是嘉荫建立了我国首个"古生物学家雕塑园";二是嘉荫国家恐龙地质公园和神州恐龙博物馆地处中俄边境,地质古生物科普工作具有明显的国际化特色;三是着力宣传了我国首个陆相K-Pg界线点(陆相"金钉子")的建立。在上述科普工作中,除恐龙作为"主导明星"外,嘉荫的古植物学研究成果宣传也独具特色。

6.1 古植物学家及古生物学家雕塑园

为纪念中外科学家对嘉荫及我国黑龙江地区地质古生物事业的贡献,嘉荫县政府于2006年在位于黑龙江畔龙骨山的嘉荫国家恐龙地质公园内,建立了我国首个

"古生物学家雕塑园"。雕塑园以长达30米的恐龙复原骨架为依托,铸立了24位国际著名科学家的青铜像,包括我国地质古生物学家李四光、杨钟健、斯行健、杨遵仪、李星学、顾知微及郝诒纯等7位院士,以及董枝明、赵喜进、孙革教授等;国外著名古生物学家包括赫胥黎(T. H. Huxley)、欧文(R. Owen)、德日进(P. T. de Chardin)、里亚宾宁(A. N. Riabinin)、克里斯托弗维奇(A. N. Kryshtofovich)、奥斯确姆(J. Ostrom)、查勒纳(B. Chalener)、迪尔切及阿克米梯耶夫等9位院士,以及玛尔凯维奇、阿什拉夫(A. R. Ashraf)、约翰森及库瑞(P. Curry)教授等(图45)。

这些栩栩如生的青铜像镌刻着嘉荫人民对科学及科学家的钟爱和崇敬,也给广大群众特别是青少年留下了深刻的记忆与影响。自2006年以来,许多国内外游客不远万里慕名前来参观,一睹嘉荫古生物学家雕塑园风采,他们在科学家的故事中,特

图45　嘉荫古生物学家雕塑园

Fig. 45　Sculpture Garden of Scientists in Jiayin

1. 雕塑园全景;2. 赫胥黎;3. 欧文;4. 德日进;5. 克里斯托弗维奇;6. 奥斯确姆;7. 李四光;8. 杨钟健;9. 斯行健;10. 杨遵仪;11. 李星学与郝诒纯(右);12. 顾知微;13. 查勒纳;14. 阿克米梯耶夫;15. 迪尔切;16. 玛尔凯维奇;17. 赵喜进;18. 董枝明;19. 孙革;20. 阿什拉夫;21. 库瑞;22. 约翰森

1. A panoramic view of the sculpture garden; 2. T. H. Huxley; 3. B. Owen; 4. P. T. de Chardin; 5. A. N. Kryshtofovich); 6. J. Ostrom; 7. Li S. G.; 8. Young C. C.; 9. H. C. Sze; 10. Yang Z. Y.; 11. Li X. X. and Hao Y. C.; 12. Gu Z. W.; 13. B. Chalener; 14. M. Akhmetiev; 15. D. L. Dilcher; 16. V. Markevich; 17. Zhao X. J.; 18. Dong Z. M.; 19. Sun G.; 20. A. R. Ashraf; 21. P. Curry; 22. K. Johnson

图46 国内外科学家在嘉荫雕塑园

Fig. 46　Int'l scientists visiting the Sculpture Garden in Jiayin

1. 中外科学家在李四光院士塑像旁;2. 董枝明教授与他的塑像;3. 俄罗斯专家在杨钟健院士(左)及孙革教授(右)塑像前;4. 印度学者和玛尔凯维奇教授(右1)在后者的塑像旁;5. 雕塑园内;6. 中外科学家在嘉荫国家恐龙公园(2019)

1. Beside the sculpture of Academician Li S. G.; 2. Prof. Dong Z. M.; 3. Sculptures of Academician Yang C. C. (left) and Prof. Sun G. (right); 4. Beside the sculpture of Prof. V. Markevich (the 1ˢᵗ right is Markevich herself); 5. In the Garden; 6. In Jiayin National Dinosaur Geopark (2019)

别是古植物学故事中,得到领悟与熏陶。科学家们的塑像也像一颗颗闪亮的明珠,将嘉荫的科学之光辉映世界(图46)。

6.2　嘉荫恐龙博物馆中的古植物化石

　　嘉荫科普工作的第二个"亮点"是建立了我国首个中俄边境地区的古生物博物馆——"嘉荫神州恐龙博物馆"和"嘉荫国家恐龙地质公园"。嘉荫恐龙博物馆内,除以黑龙江满洲龙等恐龙为"明星"作为主要展品外,还有两个高水平的植物化石展厅分别展示嘉荫晚白垩世植物化石及古新世植物群化石。

　　展厅中,展示的晚白垩世植物化石包括著名的嘉荫莲(*Nelumbo jiayinensis*)、达勒比叶(*Dalembia*)、葛赫叶(*Quereuxia*)及卡波叶(*Cabbonia*)等典型化石标本,晚白垩世恐龙与植物协同演化及其场景复原等也栩栩如生、寓教于趣(图47)。

图47　嘉荫恐龙博物馆的植物化石厅

Fig. 47　Hall of Fossil Plants in Jiayin Dinosaur Museum

1.博物馆门前；2.古植物厅；3.迪尔切院士（右2）看化石；4.李廷栋（右3）、爱德华（右4）与迪尔切（左1）院士等参观展厅；5.嘉荫莲化石；6.古生态环境复原图；7.达勒比叶；8~10.古新世乌云组剖面及植物化石

1. The front of Jiayin Dinosaur Museum; 2. Hall of Fossil Plants; 3. Dilcher（NAS，US）looking fossils; 4. Academicians Li T. D., D. Edwards and D. L. Dilcher looking fossils; 5. *Nelumbo jiayinensis*; 6. Paleo-ecological reconstruction; 7. *Dalembia*; 8–10. Section of Paleocene Wuyun Formation and its fossil plants

6.3　嘉荫地质古生物庆典活动

　　近20年来，嘉荫的古生物研究不断推向高潮，除恐龙化石的发现与研究成果尤受关注外，嘉荫的古植物学研究成果也引起科学家及广大群众的重视。这主要是由于：(1)古植物化石(特别是孢粉化石)研究准确地确定了嘉荫恐龙化石赋存的时代，并将嘉荫龙骨山和乌拉嘎两地的恐龙化石及其地层首次分开(分属于白垩纪马斯特里赫特期的早期和中期)，这也为嘉荫恐龙是"中国最晚恐龙"的论断奠定了学术基础；(2)孢粉研究首次为嘉荫K-Pg界线的确定提供了最可靠的依据，为在嘉荫建立我国陆相K-Pg界线标准(陆相"金钉子")立下了汗马功劳；(3)嘉荫莲、嘉荫达勒比叶等

众多的有花植物化石明星为嘉荫"恐龙之乡"增添了新的色彩。另一方面,嘉荫国际古生物科研团队主要由国际著名古植物学家(包括孢粉学家)组成,因此,多年来在嘉荫举行的科研与科普活动均与他们密切相关。

近20年来(2002~2019年),嘉荫先后举行了10余次有关古生物化石的大型科研、科普报告及庆典活动(图48,图49)。首次大型活动是2002年9月在嘉荫举办"首届黑龙江嘉荫白垩纪生物群与K-T界线国际学术研讨会"。中、美、德、英、俄、日、韩等多国科学家的到来,给嘉荫这座宁静的边境小城带来了喜庆的气氛和新的生机。

图48　嘉荫地质古生物庆典活动(1)

Fig. 48　Celebration activities related to Geology and Paleontology in Jiayin (1)

1~3. 首届黑龙江白垩纪生物群与K-T界线国际会议2002年在嘉荫举行,3为中央电视台报道会议召开新闻;4,5. 第三届黑龙江白垩纪生物群与K-T界线国际会议在嘉荫(2006);6. 孙革教授为嘉荫县政府机关干部作科普报告;7~9. 伊春地质古生物国际会议召开及在嘉荫举行K-Pg界线点立碑(2011)

1–3. The 1ˢᵗ int'l symposium on Cretaceous biota and KTB held in Jiayin, 2002; 3. The news footage by CCTV; 4, 5. The 3rd int'l symposium on Cretaceous biota and KTB held in Jiayin, 2006; 6. Scientific lecture of Prof. Sun G. to the staff of Jiayin Government; 7–9. Int'l symposium of Geology & Paleontology in Yichun, and setting up monument of KPgB in Jiayin, 2011

中外科学家对嘉荫古生物化石的重视,刷新了地方政府领导层和广大群众的认识。"宣传嘉荫恐龙化石、提高科学素质"以及"打造恐龙牌、带动嘉荫经济发展"等一系列新的发展理念逐渐在嘉荫滋生、开花和结果;以保护、宣传恐龙等化石为引擎的科学文化活动,也不断带动着嘉荫的经济发展和社会进步。广大群众特别是青少年,不仅在了解化石中增长了科学知识,也看到了嘉荫运用化石促进文化与经济发展的希望。2011年,嘉荫的古生物化石科普及庆典活动掀起了新的热潮:8月21~24日在嘉荫乌云小河沿举行了"伊春地质古生物国际学术研讨会"野外考察及K-Pg界线立碑庆典活动。来自世界16个国家150余名科学家出席,嘉荫又一次披上了节日的盛装。与会国内外专家对嘉荫的"K-Pg界线点"入选国际标准候选点之一(第95号点)表示由衷的祝贺,庆典取得圆满成功。

图49 嘉荫的化石庆典活动(2)

Fig. 49 Celebration activities related to Geology and Paleontology in Jiayin (2)

1、2.纪念中国恐龙发现115周年活动在嘉荫举行(2017);3.德国波恩大学师生参加活动;4~10.黑龙江嘉荫白垩纪生物群与K-Pg界线国际学术研讨会在嘉荫举行,5~7分别为刘嘉麒院士、德国莫斯布鲁格院士及美国迪尔切院士在大会上致辞(2019)

1, 2. The 115th Anniversary of Discovery of Dinosaurs in China, held in Jiayin, 2017; 3. Teachers and students of Univ. Bonn, Germany attending the celebrations; 4–10. Int'l symposium on Cretaceous biota and KPgB held in Jiayin, 2019: 5–7 showing Academicians Liu J. Q., V. Mosbrugger (Germany), and D. L. Dilcher (USA), the meeting sddressing, 2019

　　嘉荫新一轮化石宣传和科普活动的高潮是在2017~2019年。2019年9月,在中国古生物学会的大力支持下,嘉荫举行了"纪念中国恐龙发现115周年暨首届嘉荫化石保护论坛",来自全国30多个省市的专家学者以及德国波恩大学师生共100余人齐聚嘉荫。会议交流地质古生物科普经验,"嘉荫的化石保护论坛"从此成为全国的品牌交流平台。由自然资源部东北亚古生物演化重点实验室、中国古生物学会、吉林大学、沈阳师范大学及嘉荫县政府联合主办,2019年8月18~21日,黑龙江省迄今最大的一次有关化石的大型国际会议"黑龙江嘉荫白垩纪生物群及K-Pg界线国际学术研讨会暨第二届嘉荫化石保护论坛"在嘉荫举行。来自中国、美国、德国、英国、俄罗斯、比利时、罗马尼亚、日本、韩国、朝鲜、蒙古、泰国、印度、巴基斯坦、吉尔吉斯斯坦及巴西等16国的180余位专家学者出席,共同研讨嘉荫白垩纪生物群及K-Pg界线研究。与会专家高度评价我国在K-Pg界线研究中取得的新进展,并一致推荐嘉荫的"K-Pg界线点"作为我国"陆相K-Pg界线标准"。大会研讨中,嘉荫县领导也走上国际讲台,宣传嘉荫的发展,展望嘉荫的美好未来。会议期间开展的K-Pg界线科普宣传活动,被评为"2019年度全国地质古生物科普十大进展"之一。上述一系列活动的开展,不仅有力促进了嘉荫地质古生物研究成果的宣传及科学普及,也极大地推动了嘉荫的化石保护工作及地质古生物文化旅游产业的发展,使之更上一层楼(图49)。

致　谢

　　衷心感谢嘉荫国际科研队的全体专家,特别感谢科研队的古植物学团队的专家们:阿克米梯耶夫院士、迪尔切院士、高洛夫涅娃、约翰森及西田治文教授,柯珠尔及布格达耶娃研究员,以及吉林大学孙春林、孙跃武、全成教授等的辛勤奉献与协助;感谢孢粉学家玛尔凯维奇、阿什拉夫、尼科斯、哈丁以及凯金娜教授等(图50)。

图50　国际科研队的古植物学与孢粉学专家(2002~2019)
Fig. 50　Scientists of Paleobotany & Palynology in the research group in Jiayin (2002-2019)

上排左起:孙革,阿克米梯耶夫,迪尔切,玛尔凯维奇,阿什拉夫,约翰森,尼科斯,西田治文,高洛夫涅娃。下排左起:孙春林,布格达耶娃,柯珠尔,凯金娜,哈丁,孙跃武,全成,张淑芹,杨涛,梁飞
Up from left: Sun G., M. Akhmetiev, D. L. Dilcher, V. Markevich, A. R. Ashraf, K. Johnson, D. Nichols, Nishida H., and L. Goloveva. Down from left: Sun C. L., E. Bugdaeva, T. Kodrul, T. Kezina, I. Harding, Sun Y. W., Quan C., Zhang S. Q., Yang T., and Liang F.

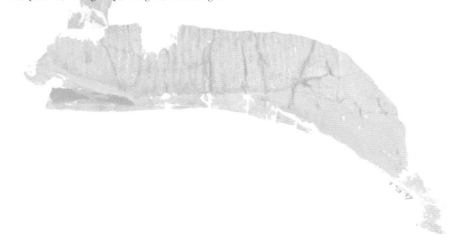

感谢恐龙学家董枝明、哥德弗洛伊特(P. Godefroit)、尤里·保罗斯基(Yu Bolotsky)、伊万·保罗斯基(I. Bolotsky),日本专家铃木茂之(Suzuki S.)、寺田和雄(Terada K.)及塚腰实(Tsukagoshi M.),俄罗斯孢粉学专家特科列娃(M. Tekleva),吉林大学杨惠心、葛文春、陈跃军教授,沈阳地调中心公繁浩及沈阳师范大学冯玉辉博士等的大力协助。感谢中国科学院李廷栋、刘嘉麒院士及王永栋教授对本书出版的支持与推荐;感谢中国地质大学(北京)万晓樵和席党鹏教授提供宝贵资料,以及上海科技教育出版社王世平总编和伍慧玲编辑对本书出版的全力支持及精心编辑工作。

感谢国家自然科学基金重大国际合作项目(3022130698)、地学部主任基金(40842002)及青年基金(41602015)项目,科技部及中国地调局"中国白垩系—古近系界线研究"项目(2015FY310100、DD20160120-04),教育部"111"项目(吉林大学,B06008)、自然资源部东北亚古生物演化重点实验室(沈阳师范大学)、东北亚生物演化与环境教育部重点实验室(吉林大学)以及现代古生物学和地层学国家重点实验室项目(183117)等资助与支持。感谢吉林大学李春田、李凌、陈峰以及沈阳师范大学古生物学院及辽宁古生物博物馆诸多同事的协助。

最后,特别感谢黑龙江省自然资源厅、黑龙江省地质博物馆、伊春市政府及嘉荫县政府、嘉荫地质公园中心及神州恐龙博物馆、黑龙江省第一地质调查所及黑龙江省第六勘察院的领导及同事们给予的大力支持与帮助。

参考文献

陈丕基.2000.中国陆相侏罗、白垩系划分对比评述.地层学杂志,24(2):114-119.

董枝明,周忠立,伍少远.2003.记黑龙江畔一鸭嘴龙足印化石.古脊椎动物学报,41(4):324-326.

董枝明.2009.亚洲恐龙.昆明:云南科技出版社,1-287.

段吉业,安素兰.2001.黑龙江伊春早寒武世西伯利亚型动物群.古生物学报,40(3):362-370.

公繁浩.2007.黑龙江嘉荫晚白垩世银杏属(*Ginkgo*)植物化石.吉林大学硕士论文,1-80.

郭双兴.1984.松辽盆地晚白垩世植物.古生物学报,23(1):85-90.

郭双兴.1986.我国及北半球白垩纪植物群面貌和演变.古生物学报,25(1):31-45.

郝诒纯,等.2000.中国地层典·白垩系.北京:地质出版社,1-124.

黑龙江省地质矿产局.1993.黑龙江省区域地质志.北京:地质出版社,1-734.

胡修棉.2004.白垩纪"温室"气候与海洋.中国地质,31:442-448.

李罡,陈丕基,万晓樵,等.2004.嫩江阶底界层型剖面研究.地层学杂志,28(4):297-299.

李锦铁,张进,杨天南,等.2009.北亚造山区南部及其毗邻地区地壳构造分区与构造演化.吉林大学学报(地球科学版),39(4):584-605.

李星学.1959.中国上白垩纪沉积中首次发现的一种被子植物——*Trapa? microphylla* Lesq.古生物学报,7(1):33-40.

梁飞,孙革.2015.黑龙江嘉荫晚白垩世永安村组水生被子植物——卡班叶(*Cobbania*)的新发现.世界地质,34(1):1-6.

梁飞,吴琪,袁亚兰,等.2018.嘉荫晚白垩世水生被子植物葛赫叶(*Quereuxia*)表皮构造.中国古生物学会第29届学术年会论文摘要集,171.

刘牧灵.1990.东北地区晚白垩世——第三纪孢粉组合序列.地层学杂志,14(4):277-285.

全成.2006.黑龙江嘉荫沿江地区晚白垩世植物群及地层.吉林大学博士研究生学位论文,1-206.

斯行健,李星学,等.1963.中国植物化石 第二册 中国中生代植物.北京:科学出版社,1-429.

孙革,曹正尧,李浩敏,等.1995.白垩纪植物群.见:李星学(主编).中国地质时期植物群.广州:广州科技出版社,310-344.

孙革,郑少林.2000.中国东北中生代地层划分对比之新见.地层学杂志,24(1):60-64.

孙革,孙春林,董枝明,等.2003.黑龙江嘉荫地区白垩纪——第三纪界线初步观察.世界地质,22(3):8-14.

孙革,全成,孙春林,等.2005.黑龙江嘉荫乌云组地层划分及时代的新认识.吉林大学学报(地球科学版),35(2):137-142.

孙革,董枝明,阿克米梯耶夫,等.2014.黑龙江嘉荫晚白垩世—古新世生物群、K-Pg界线及恐龙灭绝.上海:上海科技教育出版社,1-194.

孙革,王丽霞,孙跃武,等.2018.化石保护学基础教程.上海:上海科技教育出版社,1-159.

陶君容.2000.中国晚白垩世至新生代植物区系发展演变.北京:科学出版社,1-282.

吴文昊,Godefroit P,韩建新.2010.黑龙江嘉荫晚白垩世一鸭嘴龙亚科恐龙齿骨化石.世界地质,9(1):1-5.

吴文昊,周长付,Wing O,等.2013.新疆鄯善中侏罗世巨型蜥脚类恐龙的发现.世界地质,25(3):1-8.

席党鹏,万晓樵,李国彪,等.2019.中国白垩纪综合地层和时间框架.中国科学(地球科学),49(1):257-288.

张武,郑少林,张志诚.1980.东北地区古生物化石图册(二).北京:地质出版社,221-307.

张志诚.1981.牡丹江盆地的几种被子植物化石.中国地质科学院沈阳地质矿产所所刊,2(1):1-9.

张志诚.1984.黑龙江省北部嘉荫地区晚白垩世植物化石.地层古生物论文集,11:111-132.

张志诚.1985.东北北部白垩纪被子植物的基本发展阶段.古生物学报,24(4):453-460.

郑少林,张莹.1994.松辽盆地的白垩纪植物.古生物学报,33(6):756-764.

周志炎,李佩娟.1980.从古植物学观点讨论中国中生代陆相地层的划分、对比和时代.国际交流地质学术论文集4 地层古生物.北京:地质出版社,82-91.

Ablaev A G. 1974. *Late Cretaceous Flora of Sikhote-Alin and Its Significance for Stratigraphy*. Novosibirsk: Nauka Siberian Branch, 1-179. (in Russian)

Abramova A L. 1983. Conspectus of the moss flora of the People's Republic of Mongolia. Biological resources and natural conditions of the People's Republic of Mongolia; 17. Leningrad: Nauka, 1-221. (in Russian)

Akhmetiev M A. Biosphere crisis at the Cretaceous-Paleogene boundary. 2004. In: Sun G, Sun Y W, Akhmetiev M A, et al (eds). *Proceeding of the 3rd symposium on Cretaceous Biota and K/T boundary in Heilongjiang River area*, Changchun, 7-16.

Alvarez L W, Alvarez W, Asaro F, et al. 1980. Extraterrestrial cause for the Cretaceous-Tertiary extinction. *Science*, 208: 1095-1108.

Bell W A. 1949. Uppermost Cretaceous and Palaeocene floras of western Alberta. *Can. Geol. Surv. Bull*, 13: 1-231.

Bolotsky I. 2013. 黑龙江地区晚白垩世暴龙类恐龙(虚骨龙类).吉林大学硕士论文,1-88.

Bolotsky Y, Godefroit P. 2004. A new hadrosaurine dinosaur from the Late Cretaceous of Far Eastern Russia. *Journal of Vertebrate Paleontology*, 24: 354-368.

Dilcher D L. 1974. Approaches to the identifications of angiosperm leaf remains. *Bot. Rev.* (*Lancaster*), 40(1): 1-157.

Dilcher D L. 2000. Toward a new synthesis: Major evolutionary trends in the angiosperm fossil record. *PNAS*, 97: 7030–7036.

Doyle J A, Hickey L J. 1976. Pollen and leaves from the mid-Cretaceous Potomac Group and their bearing on early angiosperm evolution. In: Beck C B (ed). *Origin and Early Evolution of Angiosperms*. New York: Columbia University Press, 139–206.

Feng G P, Li C S, Zhilin S, et al. 2000. *Nyssidium jiayinense* sp. nov. (Cercidiphyllaceae) of the Early Tertiary from north-east China. *Bot. J. Linn. Soc.*, 134(3), 471–484.

Fontain W M. 1889. The Potomac or Younger Mesozoic flora. *US Geological Survey Monograph*, 15: 1–377.

Godefroit P, Hai S, Yu T, et al. 2008. New hadrosaurid dinosaurs from the uppermost Cretaceous of northeastern China. *Acta Palaeontologica Polonica*, 53: 47–74.

Godefroit P, Lauters P, Itterbeeck J V, et al. 2011. Recent advances on the study of hadrosaurid dinosaurs in Heilongjiang (Amur) River area between China and Russia. *Global Geology*, 13(4): 160–191.

Golovneva L B. 1994. Maastrichtian-Danian floras of Koryak Upland. *Proc. Komarov Bot. Inst. RAS.*, 13: 1–146. (in Russian with English summary)

Golovneva L B. 2005. Phytostratigraphy and Evolution of Albian-Campanian Flora in Siberia. In: *Proceedings of II All-Russia Conference on the Cretaceous System of Russia: Problems of Stratigraphy and Paleogeography* st. Petersburg. St. Petersburg: Gos. Univ., 177–197.

Golovneva L B, Sun G, Bugdaeva E. 2008. Campanian flora of the Bureya River Basin (Late Cretaceous of the Amur Region). *Paleontological Journal*, 42(5): 554–567.

Guo Z, Jia H M, Guan H T, et al. 2014. Determination of a new Late Cretaceous volcano group in Ji-anan of Liaoyuan. *Global Geology*, 33(4): 787–792.

Haq B U, Hardenbol J, Vail P R. 1987. Chronology of fluctuating sea level since the Triassic. *Science*, 235: 437–455.

He S A, Yin G, Pang Z J. 1997. Resources and prospects of *Ginkgo biloba* in China. In: Hori T, et al (eds). Ginkgo biloba, *A Global Treasure: From Biology to Medicine*. Tokyo: Springer, 373–384.

Heer O. 1878. Beitraege zur Fossilen Sibiriens und des Amurlandes. *Mem. Acad. Imp. Sci. Saint-Petersb. Ser.*, 7(25): 1–61.

Herman A B, Lebedev E L. 1991. Cretaceous stratigraphy and flora of the northwestern Kamchatka Peninsula. *Trans. Geol. Inst. Acad. Sci. USSR*, 468. (in Russian)

Herman A B. 2002. Late Early-Late Cretaceous floras of the North Pacific Region: Florogenesis and early angiosperm invasion. *Rev. Paleobot. Palyn.*, 122: 1–11.

Herman A B. 2011. Albian-Paleocene flora of the North Pacific region. *Trans. Geol. Inst.*, 592. Moscow: GEOC, 1–280.

IBP, et al. 2001. *Flora and Dinosaurs at the Cretaceous-Paleogene Boundary of Zeya-Bureya Basin*. Vladivostok: Dalnauka, 1–159.

Johnson K R. 2002. Megaflora of Hell Creek and lower Fort Union Formations in the Western Dakotas: Vegetational response to climate change, the Cretaceous-Tertiary boundary event, and rapid marine transgression. In: Hartman J H, et al (eds). *The Hell Creek Formation and the Cretaceous-*

Tertiary Boundary in the Northern Great Plains: An Integrated Continental Record of the End of the Cretaceous. Boulder: Geological Society of America, 361: 329–390.

Keller G B, Bhowmick P K, Upadhyay H, et al. 2011. Deccan volcanism linked to the Cretaceous-Tertiary boundary mass extinction: New evidence from ONGC Wells in the Krishna-Godavari Basin. *J. Geol. Soc. India*, 78: 399–428.

Knittel U, Suzuki S, Akhmetiev M A, et al. 2013. 66±1 Ma single zircon U-Pb date confirms the location of the non-marine K-Pg boundary in the Amur/Heilongjiang River area (Russia, China). *Neues Jahr. Geol. Palaont.*, 270(1): 1–11.

Krassilov V A. 1976. *Tsagajan Flora of Amur Region*. Moscow: Nauka, 1–191. (in Russian)

Krassilov V A. 1979. *Cretaceous Flora of Sakhalin*. Moscow: Nauka, 1–185. (in Russian)

Kryshtofovich A N. 1953. Some enigmatic plants of the Cretaceous flora and their phylogenetic significance. *Paleontol. Strat. Trudy VSEGEI.*, 18–30. (in Russian)

Kryshtofovich A N, Baikovskaya T N. 1966. Upper Cretaceous Tsagayan flora in the Amur Region. In: Kryshtofovich A N. Selected Papers. M.: Akad. Nauk SSSR, 3: 184–320. (in Russian)

Lebedev E L. 1974. *Albian Flora and Lower Cretaceous Stratigraphy of West Priokhotie*. Moscow: Nauka, 1–147. (in Russian)

Lebedev E, Herman A. 1989. A new genus of Cretaceous angiosperm—*Dalembia. Review of Paleobotany and Palynology*, 59: 77–91.

Lesquereux L. 1878. Contributions to the fossil flora of the Western Territories. Part 2. The Tertiary Flora. Report of the United States Geological Survey of the Territories, 7: 366.

Liang F, Sun G, Yang T, et al. 2018. *Nelumbo jiayinensis* sp. nov. from the Upper Cretaceous Yong'-ancun Formation in Jiayin of Heilongjiang, Northeast China. *Cretaceous Research*, 84: 134–140.

Markevich V S, Golovneva L B, Bugdaeva E V. 2005. Floristic Characterization of the Santonian-Campanian Deposits of the Zeya-Bureya Basin (Amur Region). In: Proceedings of International Conference on the Current Problems in Paleofloristics, Paleophytogeography, and Phytostratigraphy, Moscow, May 17–18, 2005. (in Russian)

Markevich V S, Sun G, Ashraf A R, et al. 2006. The Maastrichtian-Danian palynological assemblages from Wuyun of Jiayin nearby the Heilongjiang (Amur) River. In: Yang Q, et al (eds). Ancient life and modern appoaches. Abstracts of the 2ⁿᵈ IPC, Beijing, 526–527.

Markevich V S, Ashraf A R, Nichols D, ct al. 2008. The most important taxa for correlation of the Santonian to Danian deposits of Far East. The 2nd Workshop on the K-T boundary in Jiayin of Heilongjiang, China, Changchun, Nov. 16, 2008. 23.

Markevich V S, Bugdaeva E V, Sun G. 2009. Palynoflora of Wulaga dinosaur site in Jiayin (Zeya-Bureya Basin, China). *Global Geology*. 13(3): 117–121.

Markevich V S, Bugdaeva E V, Ashraf A R, et al. 2011. Boundary of Cretaceous and Paleogene continental deposits in Zeya-Bureya Basin, Amur (Heilongjiang) River region. *Global Geology*, 14(3): 144–159.

McLean D M. 1985. Deccan traps mantle degassing in the terminal Cretaceous marine extinctions. *Cretaceous Research*, 6: 235–259.

Miki S. 1941. On the change of flora in Eastern Asia since Tertiary Period (I). The clay or lignite beds flora in Japan with special reference to the *Pinus trifolia* beds in Central Hondo. *Jap. J. Bot.*, 11(3): 237–304.

Mosbrugger V. 1999. The nearest living relative method. In: Jones Y P, Rowe N P (eds). *Fossil Plants and Spores: Modern Techniques*. London: Geological Society, 261–265.

Mosbrugger V, Utescher T. 1997. The coexistence approach: a method for quantitative reconstructions of Tertiary terrestrial palaeoclimate data using plant fossils. *Palaeogeography, Palaeoclimatology, Palaeoecology*, 134: 61–86.

Newberry J S. 1898. Later extinct floras of North America. *US Geological Survey Monograph*, 35:1–151.

Newberry J S. 1861. Geological Report, fossil plants. In: Ives (ed). Report upon the Colorado River of the west explored in 1857 and 1858 by Lieutenant Joseph C. Ives. Corps of Topographical Engineers, Office of Explorations and Surveys. GPO, Washington, D.C., 129–132.

Nichols D, Johnson K. 2008. *Plants and the K–T Boundary*. Cambridge: Cambridge University Press, 1–292.

Quan C, Sun G. 2008. Late Cretaceous aquatic angiosperms from Jiayin, Heilongjiang, Northeast China. *Acta Geologica Sinica*, 82(6): 1133–1140.

Quan C, Sun C L, Sun Y W, et al. 2009. High resolution estimates of paleo-CO_2 levels through the Campanian (Late Cretaceous) based on *Ginkgo* cuticles. *Cretaceous Research*, 30: 424–428.

Retallack G J. 2001. A 300-million-year record of atmospheric carbon dioxide from fossil plant cuticles. *Nature*, 411: 287–290.

Riabinin A N. 1930. On the age and fauna of the dinosaur beds on the Amur River. *Mem. Russian Paleont. Soc.*, 59(2): 41–51.

Russell D. 1970. Tyrannosaurs from the Late Cretaceous of Western Canada. *Palaeontology*, 12(1): 1–34.

Samylina V A. 1963. The Mesozoic flora of the lower course of the Aldan River. *Paleobotanica*, Ⅳ. Moscow: Nauka, 59–139. (in Russian with English summary)

Samylina V A. 1967. The Mesozoic flora of the area to the west of the Kolyma River (the Zyrianka coal-basin). Ginkgoales, Coniferales. General Chapters. *Paleobotanica* Ⅵ, 133–175. (in Russian with English abstract)

Samylina V A. 1988. *Arkagalinskaya Stratoflora of Northeast Asia*. Leningrad: Nauka, 1–131 (In Russian with English abstract).

Spicer R A, Herman A B. 2001. The Albian-Cenomanian flora of the Kukpowruk River, western North Slope, Alaska: Stratigraphy, palaeofloristics, and plant communities. *Cretaceous Research*, 22 (1): 1–40.

Stockey R A, Rothwell G W, Johnson K. 2007. *Cobbania corrugata* gen. et comb. nov. (Araceae): A floating aquatic monocot from the Upper Cretaceous of western North America. *Amer. J. Bot.*, 94: 609–624.

Sun G, Zheng S L, Wang X F, et al. 2000. Subdivision of developmental stages of early angiosperms from NE China. *Act. Palaeont. Sin.*, 39 (Sup.): 186–199.

Sun G, Akhmetiev M A, Dong Z M, et al. 2002. In search of the Cretaceous-Tertiary boundary in the Heilongjiang River Area of China. *J. Geosci. Res. NE Asia*, 5(2): 105–113.

Sun G, Akhmetiev M A, Golovneva L B, et al. 2007. Late Cretaceous plants from Jiayin along Heilongjiang River, Northeast China. *Courier Forschungsinstitut Senckenberg*, 258: 75–83.

Sun G, Akhmetiev M, Markevich V, et al. 2011. Late Cretaceous biota and the Cretaceous-Paleocene (K-Pg) boundary in Jiayin of Heilongjiang, China. *Global Geology*, 14(3): 115–143.

Sun G, Golovneva L, Alekseev P, et al. 2016. New species *Dalembia jiayinensis* (Magnoliopsida) from the Upper Cretaceous Yong'ancun Formation, Heilongjiang, northern China. *Cretaceous Research*. 67: 8–15.

Sun G, Kovaleva T, Liang F, et al. 2019. A new species of *Platanus* from the Cenomanian (Upper Cretaceous) in eastern Heilongjiang, China. *Geosciences Frontiers*. 7: 8–15.

Suzuki S, Sun G, Ulrich K, et al. 2011. Radiometric Zircon Ages of a Ruff Sample from the Baishantou Member of Wuyun Formation, Jiayin: A Contribution to the Search for the K-T boundary in Heilongjiang River Area, China. *Acta Geologica Sinica*, 85(6): 1351–1358.

Sveshnikova I N. 1963. Atlas and key for the identification of the living and fossil Sciadopity-aceae and Taxodiaceae based on the structure of the leaf epiderm. *Paleobotanica* Ⅳ. Moscow: Nauka, 207–229. (in Russian with English abstract)

Taylor T N, Taylor E L, Krings M. 2009. *Paleobotany: The Biology and Evolution of Fossil Plants*. Elsevier Press, 1–1230.

Tekleva M, Markevich V, Bugdaeva E, et al. 2015. *Pseudointegricorpus clarireticulatum* (Samoilovitch) Takahashi: Morphology and ultrastructure. *Historical Biology*, 27(3–4): 355–365.

Tekleva M, Polevova S, Bugdaeva E, et al. 2019. Further interpretation of *Wodehouseia spinata* Stanley from the Late Maastrichtian of the Far East (China). *Paleontological Journal*, 53(2): 203–213.

Tekleva M, Polevova S, Bugdaeva E, et al. 2020. Three *Aquilapollenites* species from the late Maastrichtian of China: New data and comparisons. *Rev. Paleobot. Palynol.*, 282: 104288. hppts://doi.org/10.1016/j.revpalbo.2020.104288.

Vachrameev V A. 1988. *Jurassic and Cretaceous Floras and Climate of the Earth. Proceedings of the Geological Institute of the USSR, 430*. Moscow: Nauka, 1–209. (in Russian)

Wan X, Zhao J, Scott R W, et al. 2013. Late Cretaceous stratigraphy, Songliao Basin, NE China: SK1 cores. *Palaeogeography, Palaeoclimatology, Palaeoecology*, 385: 31–43.

Wang H S. 2002. Diversity of angiosperm leaf megafossils from the Dakota Formation (Cenomanian, Cretaceous), north western interior, USA. Thesis (Ph. D.). University of Florida, 1–204.

Wang H S, Dilcher D L. 2006. Angiosperm leaf megafossils from the Dakota Formation: Braun's Ranch Locality, Cloud County, Kansas, USA. *Palaeontographica* B, 273: 101–137.

Wang Y D, Huang C M, Sun B N, et al., 2014. Paleo-CO_2 variation trends and the Cretaceous greenhouse climate. *Earth-Science Reviews*, 129: 136–147.

Wolfe J A. 1995. Paleoclimatic estimates from Tertiary leaf assemblages. *Ann. Rev. Earth Planet Sci.*,

23:119−142.

Wolfe J A, Spicer R A. 1999. Fossil leaf character states: multivariate analysis. In: Jones T P, Rowe N P（eds）. *Fossil Plants and Spores: Modern Techniques*. Geological Society: London, 233−239.

Xi D P, Wan X Q, Li G B, et al. 2018. Cretaceous integrative stratigraphy and timescale of China. *Science China（Earth Sciences）*. https://doi.org./10.1007/s1430−017−9262−y.

Late Cretaceous flora from Jiayin of Heilongjiang, China

by

Ge Sun, Fei Liang, Tao Yang, and Shuqin Zhang

Preface

Jiayin is located in the northeast of Yichun region of Heilong-jiang Province, China. It's nearby the beautiful Heilongjiang (Amur) River, the border river between China and Russia. The river is immersed in a charming scenery: the dark green mountains rolling along the river; the light blue waves shining and the fishing boats rocking in the water. When you look across the river to the Russian side, the beautiful scenery is before your eyes, including the vast fields and a golden land. The vastness, beauty, and tranquility here are refreshing (Fig.1).

From the fossils of dinosaur and other organisms, we have realized that Jiayin was once a noisy "biological paradise" about 80-70 million years ago. At that time, it was warm and humid in the Jiayin area, densely covered with rivers and lakes. Dinosaurs, turtles, fish, and many other creatures lived here. In the valley and on the surrounding hillsides, the forests were dense, including many tall trees such as *Platanus*, *Metasequoia*, *Ginkgo*, and various ferns, etc. On the lake water, the lotus (*Nelumbo*) and other aquatic flowering plants grew and thrived. The Jiayin's miraculous fossils take us to the remote historic geological world, about 80-70 million years ago, the Middle and Late Cretaceous time.

Jiayin is a well-known town and honored as the "Hometown of Dinosaurs in China". In 1902, the dinosaur *Mandshurosaurus amurensis*, the first dinosaur fossil discovered in China, was unearthed in the Longgushan hill of Jiayin. And, since the dinosaurs here were the last to go extinct in China, the dinosaurs in Jiayin have the characteristics of "the earliest and the last". In recent years, with the deepening of regional geological survey and paleontological research, scientists have discovered a large number of new dinosaur fossils in Jiayin, including more than 10 taxa such as *Mandshurosaurus*, *Charonosaurus*,

Wulagasaurus, *Sahaliyania*, *Amurosaurus*, *Olorotitan*, *Kundursaurus*, *Kerbersaurus*, *Jiayinosau-ropis*, and Tyrannosaurid. They provide the most ideal materials for studying the evolution of di-nosaurs in the Late Cretaceous in China and even Northeast Asia. At the same time, a large number of plant fossils have also been found in the Upper Cretaceous in Jiayin, especially the aquatic angiosperm *Nelumbo jiayinensis*, which adds new luster to the fossil land of Jiayin. Meanwhile, with the request of stratigraphic and paleoenvironmental correlation between the eastern and western Heilongjiang Province, the studies on the associated biota of Jiayin dino-saurs and their paleogeography and paleoclimate have also been put on the agenda. In recent years, the studies on dinosaur fossils show that about 95% of the dinosaurs in Jiayin belong to Hadrosaurid in taxonomy, while the flourishing plants in the Jiayin area were the food sources for the dinosaurs survival and development during the Late Cretaceous. Thus, how about the composition of the Late Cretaceous flora in Jiayin? Which plants had been on dinosaur's "din-ing table"? Did the rise and fall of these plants affect the development and extinction of the di-nosaurs? How did dinosaurs and plants coevolve? All the scientific problems need to be faced by geoscientists, especially by the paleobotanists.

Therefore, since 2002, an international scientific research team led by Chinese paleobota-nist Sun G. and joined by the experts from Russia, Germany, USA, UK, Belgium, Japan, and China, has carried out the research project work in the Jiayin area for more than 10 years (Fig. 2), and gained important achievements in the study of the Cretaceous biota and the K-Pg boundary, including on the dinosaurs, plant fossils and quantitative study of paleoclimate by *Ginkgo* fos-sils. Paleobotanically, the new achievements are highlighted by the discoveries of the Late Cre-taceous important angiosperms *Dalembia*, and a large number of aquatic angiosperm fossils in Jiayin, such as *Nelumbo*, *Quereuxia*, and *Cabbonia*, which not only further revealed the composi-tion of the Late Cretaceous flora in Jiayin area, but also achieved the comprehensive study of pa-laeogeography and paleoclimate of Jiayin and its adjacent areas in the Late Cretaceous, and the evolutional sequence of the Late Cretaceous floras in China. Since 2015, the authors have un-dertaken two important scientific research projects on the "Cretaceous-Paleogene bound-ary in China" financed by the Ministry of Science & Technology, China, and China Geological Survey, which has gained new achievements also. Therefore, a comprehensive summary and publishing the report on the latest research achievements on the Late Cretaceous flora in Jiayin, are taken for granted by the authors.

Historically, the study of the Late Cretaceous flora of Jiayin was first made by Prof. Zhang Z. C., paleobotanist in Shenyang Geological Survey Center (formerly the SIGMR). After that,

Profs. Guo S. X. (NIGPAS) and Tao J. R. (IB, CAS) and other paleobotanists have made their contributions to the related researches. For the past 20 years since 2002, the international scientific research team led by Sun G. (JU/ SYNU) has made great contributions to the study of the Late Cretaceous flora and its related research in Jiayin. The research team includes the experts Profs. M. Akhmetiev, L. Golovneva, and V. Markevich (Russia), D. L. Dilcher (NAS), K. Johnson and D. Nichols (USA), A. R. Ashraf (Germany), Nishida H. and Suzuki S.(Japan), as well as Chinese experts Profs. Sun C. L., Sun Y. W., Quan C., Chen Y. J., Yang H. X., Ge W. C., and Dr. Feng Y. H., etc. Therefore, many results of this book are as a crystallization of collective wisdom and sweat of all the experts in the research team.

The Late Cretaceous (ca. 100.5-66.0 Ma), the last period of the Mesozoic, is one of the time periods that witnessed drastic changes in the earth's history. In this period, dinosaurs eventually went extinct, and the global vegetation also changed greatly. The publication of this book is of great significance to the study of Mesozoic geo-/biological evolution history of China and even the Northeast Asia, and also provides an important reference for the study of the Cretaceous floras in the world.

This book consists of 6 chapters with 18 sections. For the first time it has comprehensively reported the composition, characteristics, age, palaeogeography, and paleoenvironment of the middle-late Late Cretaceous flora in Jiayin, China. Besides the fossil plants of 34 genera and 43 species described in the book, the co-evolution of the Late Cretaceous flora with dinosaurs in Jiayin, and the related K-Pg boundary, and the current status of the study of the evolutionary sequence of the Late Cretaceous floras in Northeast China, all are briefly introduced. It is appreciated that the book has told the stories of "the dining table for herbivory dinosaurs" and "the paleontologist sculpture garden in Jiayin" for science popularization for the first time. Thus, it seems to be that this book is the first systematically scientific monograph on the Late Cretaceous flora of Jiayin, and a popular science reading material closely related to study of dinosaurs in Jiayin, for readers.

June 2020

Nelumbo jiayinensis

Chapter **1**

Outline of Geology in Jiayin area, Heilongjiang

1.1 Outline of Geology

Jiayin is located in the northern Yichun region of Heilongjiang Province (central coordinates: 130°32′44″E, 48°53′29″N; Fig. 3, 1). Yichun region is located in the composite orogenic belt (Xingmeng orogenic belt) between the Siberian Plate and North China Plate, and the intersection of the eastern Paleo-Asian Ocean Tectonic Domain with the Coastal Pacific Tectonic Domain (Li et al., 2009). Geologically, the earliest records in Yichun region is the Early Cambrian (more than 500 Ma). In the Cambrian and Middle Ordovician seas were covering the Yichun region, evidenced by trilobite *Proerbia* found in Wuxingzhen Formation of Xilin, Yichun (Duan & An, 2001), and the marine fossils of Early and Middle Ordovician found in the Duobaoshan area adjacent to Yichun (BGMRHP, 1993). In the Middle Ordovician (ca. 470-460 Ma), the Yichun region was influenced by the Middle Caledonian movement, causing a large-scale volcanic and magmatic activities, and forming a lot of metal minerals. Since the Early Devonian, the crust in Yichun was uplifted and absent in strata. In the Early-Middle Devonian (ca. 410 Ma), the Yichun region was covered by coastal and shallow sea. During the Late Devonian to Carboniferous, the region was uplifted again. In the Middle-Late Permian (ca. 270-252 Ma), the Yichun region was mainly continental in deposits, with alternative marine and continental deposition in some parts. The famous Angara flora was found in the Upper Permian of Hongshan in Yichun. By the end of the Late Permian, violent volcanic activity appeared again in the Yichun region.

In the Mesozoic (ca. 252-66 Ma), the Yichun region was completely occupied by land. In the Triassic and Jurassic, this region was distributed with granites, volcanic rocks, and their sedimentary clastic rocks, caused by the frequent collision between the Pacific Plate and the Eurasian Continental Plate. In the Cretaceous, the Jiayin basin was a part of the Zeya-Bureya basin of Russia (Fig. 3, 2), and mainly deposited by the Middle-Upper Cretaceous strata (ca. 88-66 Ma), including Yong'ancun Formation (fluvial facies with lacustrine), Taipinglinchang Formation (lacustrine facies), Yuliangzi Formation (fluvial facies), and Furao Formation (lacustrine facies), in ascending order. A large number of dinosaur fossils are found in the coarse clastic rocks of the Yong'ancun and Yuliangzi formations. The Furao Formation comforms with the Paleocene Baishantou Member of Wuyun Formation, documented by the boreholes nearby the village Xiaoheyan of Wuyun, Jiayin. The sedimentary strata underlying the conformably Yong'ancun Formation are absent. While, in Zhaojiadian (48°44′49.1″N, 130°33′05.5″E), south of the Yong'ancun, the authors have found a set of acidic volcanic rocks with U-Pb isotopic dating as 101.6 ± 1.2 Ma, recently, which means that there might be an absence of the Lower and lower Middle Cretaceous in Jiayin, and the volcanic strata dated as the Albian, underlying unconformably the Yong'ancun Formation, perhaps belongs to the Ningyuancun Formation, as mentioned by BGMRHP (1993).

Since the Paleocene (ca. 66 Ma) in the Jiayin basin, the fluvial and lacustrine facies sediments were deposited with a thick brown coal-bearing strata forming the Wuyun Coal Mine, which was mainly caused by the expanding of the Zeya-Bureya basin and the development of the new large-scale fault depression in the northeastern Heilongjiang Province. After that, with the neotectonic movements, the Heilongjiang valley was formed, with the sediments of mainly fluvial facies. In the Quaternary, the crustal movement is characterized by differential rise and fall, and after many glacial activities, the land gradually flattened and the climate gradually turned warm in the Jiayin area, which is close to today's appearance.

1.2 Upper Cretaceous strata

The Upper Cretaceous in the Jiayin area is well developed, mainly including Yong'ancun-, Taipinglinchang-, Yuliangzi- and Furuo formations (Table 5). The mega-fossil plants reported in this book are mainly collected from the Yong'ancun Formation and Taipinglinchang Formation. Dinosaur fossils are mainly found in the Yuliangzi Formation, but the dinosaur track was found in Yong'ancun Formation (Dong et al., 2003). Since the Jiayin basin belongs to the Zeya-Bureya basin, dominated in the Russian side, the Upper Cretaceous and Paleocene are also widely developed in the Russian areas, highlighted mainly by Belay Gora, Akhara, and other places on the left bank area of the Heilongjiang (Amur) River (Fig. 1), which provides good conditions for the study of correlation on the Cretaceous stratigraphy and the Cretaceous-Paleogene (K-Pg) boundary between the Jiayin and Zeya-Bureya basins (Sun et al., 2014).

Table 5 Stratigraphy of Upper Cretaceous in Jiayin of Heilongjiang and correlation with its adjacent area in Russia (after Sun et al.,2014,with revision)

Formation / Region / Stage			Russia Zeya-Bureya Basin				China Jiayin of Heilongjiang	
Paleocene	Danian	U	Tsagayan Group	U	Upper Tsagayan Fm.	Coal-bearing beds	Wuyun Fm.	Coal-bearing Member
Paleocene	Danian	L	Tsagayan Group	U	Upper Tsagayan Fm.	Sandstone beds	Wuyun Fm.	Baishantou Member
Upper Cretaceous	Maastrichtian	U	Tsagayan Group	M	Bureya Fm.		Furao Fm.	
Upper Cretaceous	Maastrichtian	M	Tsagayan Group	L	Udurchukan Fm. (Main dinosaur horizon)		Yuliangzi Fm. (Main dinosaur horizon)	U
Upper Cretaceous	Maastrichtian	L	Tsagayan Group	L	Udurchukan Fm. (Main dinosaur horizon)		Yuliangzi Fm. (Main dinosaur horizon)	L
Upper Cretaceous	Campanian		Kundur Fm.		Upper Zavitin Fm.		Taipinglinchang Fm.	
Upper Cretaceous	Santonian		Kundur Fm.		Upper Zavitin Fm.		Yong'ancun Fm.	
Upper Cretaceous	Coniacian		Boguchan Fm.		Upper Zavitin Fm.		?	
Upper Cretaceous	Turonian		Boguchan Fm.		Upper Zavitin Fm.		?	
Upper Cretaceous	Cenomanian		Lower Zavitin Fm.				?	

1.2.1 Yong'ancun Formation (K$_{2yn}$)

The Yong'ancun Formation is mainly exposed in the east hill of Yong'ancun village to the east of Jiayin Town. The formation is mainly composed of yellow gray sandstone, siltstone, mudstone, sandy conglomerate lens, with thin layers of gypsum, few coal seams, and thin tuff at the top, showing fluvial and lacustrine facies. The thickness of the formation is more than 125 m. The typical section of this formation is exposed in the east hill of Yong'ancun village (central coordinates: 130°31′24″E, 48°50′57″N). Overlying the formation in conformity is an oil-shale at the bottom of the Taipinglinchang Formation. The oil-shale yielding conchostraca *Halysestheria yui* (Chang), is considered to be equivalent to the bottom of the Nenjiang Formation in Songliao basin (Li et al., 2004). Underlying the Yong'ancun Formation in unconformity is mainly acid volcanic rocks which are considered to be the Ningyuancun Formation of the Lower Cretaceous (BGMRHP, 1993). Not long ago, the authors found a set of acid volcanic rocks in Zhaojiadian (130°33′05.5″E, 48°44′49.1″N) in the south of the Yong'ancun, with the U-Pb dating the volcanic rocks as 101.6 ± 1.2 Ma in age[*], which appears to document the volcanic rocks might be the Lower Cretaceous Ningyuancun Formation.

The Yong'ancun Formation richly yields plant fossils, with some animal fossils including ostracods, bivalves, and dinosaur track *Jiayinosaurupus johnsoni* (Dong et al., 2003) (Fig. 4, 4). The plant fossils are composed of abundant mega- and micro- fossils. The megafossil plants are characterized by *Parataxodium - Nelumbo* Assemblage, composed of over 27 species of 24 genera, including ferns: *Equisetum* sp., *Asplenium dicksonianum* Heer, *Arctopteris?* sp., *Cladophlebis* sp., *Gleichenites* sp.; ginkgoales: *Ginkgo adiantoides* (Ung.) Heer, *G. pilifera* Samylina; conifers: *Parataxodium* sp., *Metasequoia disticha* (Heer) Miki, *Sequoia* sp., *Cupressinocladus sveshnikovae* Ablajev, *Elatocladus* sp. 2; angiosperms: *Dalembia jiayinensis* Sun et Golovneva, *Trochodendroides arctica* (Heer) Berry, *Nyssidium articum* (Heer) Iljinskaja, *Platanus* sp., *Cobbania corrugata* (Lesq.) Stockey et al., *Nelumbo jiayinensis* Liang et al., *Quereuxia angulata* (Newb.) Krysht., etc. (Sun et al., 2014; Liang & Sun, 2015; Sun et al., 2016; Liang et al., 2018a). The microfossil plants are sporopollen

---

[*] Sun G. Annual report of the project "The study of Cretaceous Paleogene boundary in China" (MOST, Proj. 2015FY310100) (unpublished data). 2017.

composed of *Kuprianipollis santaloides-Duplosporis borealis* Assemblage. The age of the Formation is considered as the Santonian of the Late Cretaceous (Markevich et al., 2011).

1.2.2 Taipinglinchang Formation (K$_{2tp}$)

The Taipinglinchang Formation is mainly exposed in Taipinglinchang of Jiayin, on the right bank of Heilongjiang River (central coordinates: 130°14′58″E, 48°51′31″N). The formation is mainly composed of yellowish-greenish gray sandstone, siltstone, mudstone, and oil-shale, showing mainly lacustrine facies in deposition. The formation is in parallel unconformity with the overlying Yuliangzi Formation; and the oil shale at the bottom is in conformity with the underlying Yong'ancun Formation. The total thickness of the formation is more than 635 m (Fig. 5). Since the oil-shale yields conchostraca *Halysestheria yui* (Chang) (= *Halysestheria qinggangensis* Zhang et Chen) which is the index-fossil of the Nenjiang Formation, the Taipinglinchang Formation can be correlated with the Nenjiang Formation of the Upper Cretaceous in Songliao basin, aged in the Campanian. Beside, the authors found the fossil fish, *Sungarichthys longicephalus* Takai, from the Taipinglinchang Formation in the Target Range No. 4 of Jiayin (Figs. 5-5, 6). This fish taxon was found from the Fulongquan Formation (equals to Nenjiang Formation) of Jilin, which provided an important evidence for the correlation of the Taipinglinchang Formation of Jiayin with the Nenjiang Formation in Songliao basin.

The Taipinglinchang Formation is rich in plant fossils whose assemblage was named as *Metasequoia - Trochodendroides - Quereuxia* Assemblage (Sun et al., 2007, 2011; Sun et al., 2014). Up to now, 38 species of 30 genera of the megafossil plants have been reported, including bryophytes: *Thallites jiayinenssis* Zhang; ferns: *Equisetum* sp., *Asplenium dicksonianum* Heer, *Cladophlebis* sp.; ginkgoales: *Ginkgo adiantoides* (Ung.) Heer, *G. pilifera* Samylina; conifers: *Parataxodium* sp., *Metasequoia disticha* (Heer) Miki, *Sequoia* sp., *Larix* sp., *Gryptostrobus* sp., *Pityophyllum* sp., *Pityospermum* sp., *Elatocladus* spp. 1, 2; angiosperms: *Araliaephyllum?* sp., *Arthollia tschernyschewii* Golovneva, Sun et Bugdaeva, *A. orientalia* (Zhang) Gol., *Celastrinites kundurensis* (Konstanov) Gol., Sun et Bugd., *Platanus multinervis* Zhang, *P. sinensis* Zhang, *Platanus* sp., *Trochodendroides arctica* (Heer) Berry, *T. lanceolata* Golovneva, *T. smilacifolia* Zhang, *T. taipinglinchanica* Golovneva, Sun et Bugdaeva, *Viburnophyllum* sp., *Quereuxia angulata* (Newb.) Krysht., *Cobbania corrugata*

(Lesq.) Stockey et al. (Sun et al., 2007, 2011, 2014). The sporopollen fossils are composed of the Assemblage of *Aquilapollenites conatus - Podocarpidites multesimus*, showing the age of the Late Cretaceous Campanian (Markevich et al., 2006, 2011; Sun et al., 2014).

1.2.3 Yuliangzi Formation (K_{2yl})

The Yuliangzi Formation is mainly distributed in the Longgushan hill, east of Yuliangzi village, on the right bank of Heilongjiang River (central coordinates: 130°13′40″E, 48°51′30″ N), and the Wulaga area in south of Jiayin. The formation is divided into upper and lower members in which the lower member is mainly exposed in the Longgushan hill, composed of grayish-green and grayish-yellow conglomerate, gravelly sandstone, sandstone, and mudstone, with a thickness of 287 m. This member yields a large number of dinosaurs highlighted by *Mandschurosaurus*. The sporopollen in this member are represented by *Aquilapollenites amygdaloides - Gnetaceaepollenites evidens* Assemblage (Assemblage Ⅲ), indicating the early Maastrichtian age. The upper member of the formation is mainly exposed in the Wulaga area, consisting of a set of grayish purple conglomerate and sandstone, yielding dinosaurs also, such as *Wulagasaurus dongi* God. et al., *Salaliyania elunchunorum* God. et al., etc. (Godefroit et al., 2011) (Fig. 6). The sporopollen fossils from the upper member represented by the assemblage of *Wodehouseia aspera - Parviprojectus amurensis* (Assemblage Ⅳ), indicating the middle Maastrichtian age (Markevich et al., 2006, 2011; Sun et al., 2014).

Oblique bedding and cross-bedding are widely developed in the Yuliangzi Formation, which clearly shows the fluvial facies. The contact relation between the formation and the overlying Furao Formation is unknown (Fig. 6). The Yuliangzi Formation is very rich in dinosaurs, which provides strong evidence for the age of the formation, e.g. *Albertosaurus periculosus* Riabinin in the formation, was found in the Upper Cretaceous Maastrichtian of Alberta, Canada (Russell, 1970). Some experts on dinosaur have considered the dinosaur fauna of *Mandschurosaurus-Albertosaurus* as the middle and late Maastrichtian age. However, according to the palynological study, the dinosaur-bearing beds in Jiayin are considered as the early-middle Maastrichtian (Markevich et al., 2006, 2011; Sun et al., 2011, 2014).

1.2.4 Furao Formation（K_{2fr}）

The Furao Formation is the uppermost formation of the Upper Cretaceous in Jiayin, and one of the most important strata for the study of the K-Pg boundary in the Jiayin area. The formation is mainly composed of dark gray mudstone, fine siltstone, with tuffaceous siltstone locally, which is mainly found in the boreholes at the Xiaoheyan village in Wuyun, Jiayin, representing a set of lacustrine deposits. The thickness of the formation is more than 136 m (Sun et al., 2014). The Furao Formation was named by the Geological Survey No.1 of Heilongjiang in 1981, with the original meaning as a set of fluvial-lacustrine facies deposits, mainly composed of dark gray siltstone, sandstone, tuffaceous breccia, and acid tuff, with the age excessive from Cretaceous to Tertiary (BGMRHP, 1993), based on the palynological study (Liu, 1990). However, since 2002, according to the new study carried out by the international research team led by Sun G., in the upper part of the previous "Furao Formation", some Paleocene fossils (e.g. *Tiliaphyllum tsagajanense*, etc.) were discovered, and the isotopic dating of the acid tuff from the upper part, showing the Daning age. Therefore, the new study has divided the previous "Furao Formation" into the two parts that the previous upper part (including the acid tuff, dark brown fine siltstone with thin coal seams) is renamed as "Baishantou Member" belonging to the lower member of the Wuyun Formation aged as Paleocene, while the middle and lower parts of the previous "Furao Formation" (composed of a set of siltstones, mudstones, and sandstones, etc.) are revised as the present Furao Formation, aged as the late Maastrichtian (Sun et al., 2002, 2005, 2014).

However, as the newly divided Furao Formation has not been fully exposed on the surface, Sun et al. drilled three boreholes in the south of Xiaoheyan village during 2005-2008 (XHY-2005, 2006, 2008, with central coordinates of 129°35′14″E, 49°14′53″N), revealing the composition of the new Furao Formation successfully and making the definition of the K-Pg boundary in the XHY-2006 at 22.00-22.05 m(Sun et al., 2011, 2014)(Fig. 7).

Although no well-preserved megafossil plants have been found in the Furao Formation, there are abundant sporopollen fossils found and studied in detail. The palynoflora of the Furao Formation is represented by the assemblage of *Aquilapollenites conatus-Pseudoaquila-Pollenites striatus* (Assemblage V), composed of 49 species of 29 genera, in which there are the Late Cretaceous typical taxa including *Marsypiletes cretacea, Tricolpites variexinus, Aquilapollenites stelkii, A. proceros, A. striatus, A. rigidus, Intergricorpus bellum, Pseu-*

dointegricorpus clarireticulatus found, and the taxa *Aquilapollenites stelkii*, *A. conatus*, *Pseudointegricorpus clarireticulatus*, *Marsypiletes cretacea*, *Integricorpus bellum* are limited to the late Maastrichtian in age. Thus, the sporopollen indicate in detail the Furao Formation as the late Maastrichtian age (Markevich et al., 2006, 2009, 2011; Sun et al., 2014).

Chapter 2

Late Cretaceous flora in Jiayin

2.1 Composition of flora

In the Late Cretaceous (100.5-66.0 Ma), the global climates were generally hot and dry. However, in the Pacific coast region, e.g. the Far East of Russia and the eastern Heilong- jiang of Northeast China, due to its short distance from the sea, and being affected by the warm and humid climates, the vegetation was still flourishing (Herman, 2004), and its lush forests looked similar to the low mountains and warm temperate zone in the present Yangtze River region of China (Sun et al., 2014). According to a large number of plant fossils found in Jiayin, the middle-late Late Cretaceous (Santonian-Campanian, ca. 86-72 Ma) flora was growing abundantly and diversely, in which the moss, ferns, ginkgos, coni- fers, and angiosperms were main taxa in the lush forests. The flora in Jiayin is highlighted by the flowering plants which were becoming the dominant group (Sun et al., 2007, 2014).

Compared with the Early Cretaceous flora in composition, the Late Cretaceous flora in Jiayin is greatly reduced in ferns, cycads, even ginkgos, and nearly disappeared in the cze- kanowskias. No cycads have been found in the middle-late Late Cretaceous flora of Jiayin, so far. In general, the conifers still account for a large proportion (ca. 27%), in which some new taxa, such as *Metasequoia*, *Sequoia*, and *Gryptostrobus*, appeared. The highlight is that the angiosperms were greatly developed and suddenly increased in diversity, such as *Trochodendroides*, *Celastrinites*, *Dalembia*, *Platanus*, and a large number of aquatic angiosperms (e.g. *Quereuxia*, *Cobbania*, and *Nelumbo*) occurred, and many angiosperms were close to their related living taxa (Zhang, 1984; Quan, 2006; Sun et al., 2007, 2011, 2016; Golovneva et al., 2008; Liang & Sun, 2015; Liang et al., 2018a).

Up to now, over 43 species of 34 genera have been found in the Jiayin middle-late Late Cretaceous flora, including bryophytes (1 taxon), ferns (5 taxa), ginkgos (3 taxa), conifers (12 taxa), and angiosperms of 21 species belonging to 15 genera with the ration about 50.0% in the floral composition, clearly dominant in the flora.

2.1.1 Mega-fossil plants

The middle-late Late Cretaceous flora of Jiayin can be divided into two assemblages in general, including: ① the Yong'ancun Formation Assemblage (*Parataxodium-Nelumbo* Assemblage) with Santonian age; ② the Taipinglinchang Formation Assemblage (*Metasequoia-Trochodendroides-Quereuxia* Assemblage), aged as the Campanian (Sun et al., 2007, 2014). The list of the taxa and their geological horizons are shown in Table 2 (pages 19, 20).

① The Yong'ancun Formation Assemblage (*Parataxodium-Nelumbo* Ass.)

This assemblage is mainly yielded in the Yong'ancun Formation. Up to now, more than 27 species of 24 genera have been found, mainly including ferns: *Equisetum* sp., *Asplenium dicksonianum* Heer, *Arctopteris* sp., *Cladophlebis* sp., *Gleichenites* sp.; ginkgo: *Ginkgo adiantoides* (Ung.) Heer, *G. pilifera* Samylina; conifers: *Cupressinocladus sveshnikovae* Ablajev., *Metasequoia disticha* (Heer) Miki, *Sequoia* sp., *Parataxodium* sp., *Elatocladus* sp. 2; angiosperm: *Dalembia jiayinensis* Sun et Golovneva, *Menispermites* sp., *Trochodendroides arctica* (Heer) Berry, *Nyssidium arcticum* (Heer) Iljinskaja, *Platanus* sp., *Viburnophyllum* sp., *Dicotylophyllum* sp., *Quereuxia angulata* (Newb.) Krysht., *Cobbania corrugata* (Lesq.) Stockey et al., *Nelumbo jiayinensis* Liang et al. (Sun et al., 2014, 2016; Liang, 2015; Liang et al., 2018a). According to the sporopollen fossils, the age is the Late Cretaceous Santonian (Markevich et al., 2011; Sun et al., 2014).

② The Taipinglinchang Formation Assemblage (*Metasequoia-Trochodendroides-Quereuxia* Ass.)

This assemblage is mainly yielded in the Taipinglinchang Formation. More than 38 species of 30 genera have been found, mainly including bryophytes: *Thallites* sp.; ferns: *Equisetum* sp., *Asplenium dicksonianum* Heer, *Cladophlebis* sp.; ginkgos: *Ginkgo adiantoides* (Ung.) Heer, *G. pilifera* Samylina; conifers: *Taxodium olrikii* (Heer) Brown, *Metasequoia disticha* (Heer) Miki, *Sequoia* sp., *Pityophyllum* sp., *Pityospermum* sp., *Glyptostrobus* sp., *Larix* sp., *Elatocladus* sp. 1, 2; angiosperm: *Araliaephyllum?* sp., *Arthollia orientalis* (Zhang) Golovneva, *A. tschernyschewii* (Kostanov) Golovneva, Sun et Bugdaeva, *Celastrinites kundurensis* Gol., Sun et Bugd., *Platanus densinervis* Zhang, *P. sinensis* Zhang, *Platanus* sp., *Trochodendroides arctica* (Heer) Berry, *T. taipinglinchanica* Gol., Sun et Bugd., *T. lanceolata* Gol., *T. microdentaus* (Newb.) Krysht., *Viburnum* cf. *contortum* Berry, *Viburnophyllum* sp., *Quereuxia angulata* (Newb.) Krysht., *Cobbania corrugata* (Lesq.) Stockey et al. (Sun et al., 2007, 2011, 2014; Golovneva et al., 2008). According to the sporopollen fossils, the age is the Late Cretaceous Campanian (Markevich et al., 2011; Sun et al., 2014).

Systematics

Bryophytes

Thallites sp. (Fig. 8, 1)

Moss-shaped leaves, generally wedge-shaped; leaves small, dichotomous for 2-3 times; lobes long linear, each lobe about 4-5 mm long by 1-1.5 mm wide, blunt round apex and flat surface with a single midvein. Reproductive organs not preserved.

Thallites jiayinensis described by Zhang (1984) is actually the submerged leaves of *Quereuxia*, and corrected by Quan, and Sun (Quan, 2006; Sun et al., 2014).

Horizon: Upper Cretaceous, Taipinglinchang Formation.

Filiales

Equisetales

Equisetum sp. (Fig. 8, 2, 3)

Stem incomplete, with sheathed leaves in preservation. Stem thin, about 0.8-1.0 cm wide, noded by a leaf sheath, 1.2-1.5 cm long, with about 10 leaves. Leaf about 1-2 mm wide, with the free part of 4-6 mm long, acuminate in apex. Reproductive part not preserved.

Horizon: Upper Cretaceous, Yong'ancun and Taipinglinchang formations.

Filicales

Asplenium dicksonianum Heer (Fig. 8, 4-7)

Pinnae are incomplete in preservation. The ultimate secondary pinnule nearly lanceolate or rhombus, about 3-4 cm long by 1.5-2 cm wide. Rachis thin, about 1 mm wide. Pinnules nearly lanceolate or rhombus in form, subalternate, about 1.5-2 cm long by 4-5 mm wide, slightly acute in apex. Reproductive part not preserved.

Horizon: Upper Cretaceous, Yong'ancun and Taipinglinchang formations.

Cladophlebis sp. (Fig. 8, 8-9)

Osmundaceous-like ferns. Pinna nearly linear or long lanceolate in form, over 8 cm long by about 2 cm wide, with thin rachis, about 1 mm wide. The pinnules nearly elliptic-shaped, subalternate, 0.8-1 cm long by 4-5 mm wide for each, obtuse in apex. Midvein straight, lateral veins branched for 1-2 times. Reproductive parts not preserved.

Horizon: Upper Cretaceous, Yong'ancun and Taipinglinchang formations.

Gleichenites sp. (Fig. 8, 10-11)

Frond small in size, with relatively coarse axis. Pinnae nearly linear to elongate lanceolate, about 0.8-1.2 cm long by about 0.3 cm wide, for each. Pinnules slightly subbroadly elliptic in form, about 2-3 mm long by 1-1.5 mm wide for each, obtuse in apex, subopposite, densely arranged at axis. Veins pinnate, not very clear. Reproductive parts not preserved.

Horizon: Upper Cretaceous, Yong'ancun Formation.

Gymnospermae

Ginkgoales

Ginkgoales, as an important part of the flora, possesses 3 species of 1 genus (*Ginkgo*), including *G. adiantoides* (Ung.) Heer, *G. pilifera* Samylina and *Ginkgo* sp., found in the Late Cretaceous flora of Jiayin. Although the diversity of *Ginkgo* is not higher, its quantity is larger, and distributions are wide, ranging from the Yong'ancun to the Taipinglinchang Formation, usually well preserved in cuticles. The study of these fossil cuticles of *Ginkgo* is important for the classification of ginkgos, and reconstructions of the climate and environment of the Late Cretaceous in the Jiayin area (Quan, 2006; Gong, 2007; Sun et al., 2007; Quan & Sun, 2008; Quan et al., 2009; Sun et al., 2014) (Fig. 9).

Ginkgo adiantoides (Ung.) Heer (Fig. 9, 1-8; Fig. 10)

Lamina semi-circular or fan-shaped, 3-4 cm long by 5-6 cm wide; margin entire or slightly undulate; base broadly cuneate, veins extending from the base, dichotomous.

Cuticle hypostomatic. In upper cuticles, the coastal and intercoastal zones well defined. Coastal zone usually consisting of 3-5 files of cells. Ordinary epidermal cells elongated, about 71-122 μm × 12-18 μm in size. Anticlinal walls straight or undulate; and periclinal walls usually having a papilla for each. In intercoastal zones, ordinary epidermal cells irregular quadrilateral or polygonal in form, about 32-50 μm × 18-35 μm in size, anticlinal wall undulated curve, periclinal wall flat, occasionally with a papilla for each.

In lower cuticles, coastal zone usually consisting of 4-6 files of elongated cells; anticlinal walls straight and slightly undulate; periclinal walls with a papilla for each, occasionally hairy. The epidermal cells in the interveinal region not obvious in form. Stomata cyclocytic, about 17-26 μm × 10-13 μm in size for each, random in orientation; the guard cells sunken and thickened in inner edge; with 4-7 subsidiary cells strongly papillate, covering partly apertures (Fig. 10, 2-8).

The present materials are basically consistent in gross morphology and cuticular characters with this species from the Siberia described by Samylina (1963, 1967).

Horizon: Upper Cretaceous, Yong'ancun and Taipinglinchang formations.

Ginkgo pilifera Samylina (Fig. 9, 9-10; Fig. 11)

Leaves nearly complete in preservation, semicircular to fan-shaped, about 3 cm long by 6 cm wide, entire or slightly undulate, occasionally notched in central apex; with a longer stalk, over 2.5 cm long by about 2 mm wide. Veins dichotomously branched from the base. The cuticles well preserved, and mainly found in the lower epidermis, occasionally in the upper epidermis. A large number of scattered trichomes developed in the upper and lower epidermis. In the upper cuticles, the ordinary epidermal cells about 83-121 μm × 13-21 μm or 31-56 μm × 18-31 μm in size for each. The anticlinal walls straight or wavy curved, and the periclinal wall uneven. Trichome spiky conical in form, up to 40 μm long, with a base diameter of about 16 μm, scattered in distribution. Stomata occasionally seen.

In lower cuticles, the ordinary epidermal cells with straight or slightly curved anticlinal walls; and the periclinal walls papillate with trichomes. The trichome conical, up to 45 μm long with the base about 18 μm in diameter (Fig. 11, 9). Stomata cyclocytic, about

19-26 μm × 18-23 μm in size for each, random in orientation. Guard cells radially cuticularized with thickened striations at the inner edges (Fig. 11, 7). Subsidiary cells 5-6 in number, strongly papillate, almost covering the apertures (Fig. 11, 8).

The present material is basically consistent with the type specimens from the Upper Cretaceous of Russian Siberia (Samylina, 1967), while is different only in that the ordinary epidermal cells of Russian material are centrally thickening or papillate.

Horizon: Upper Cretaceous, Yong'ancun and Taipinglinchang formations.

Ginkgo sp. (Fig. 9, 11-14; Fig. 12)

Leaves fan to semicircular, around 1.9-3.1 cm long by 3.2-4.4 cm wide; entire, or microwave, or lobed once in the middle apex of the lamina. Stalk about 1.2 cm long by 1.5 mm wide, and 13-15 veins dichotomous stretching from the leaf base.

Cuticle amphistomatal. In the upper cuticles, the stomata few in number, and unevenly distributed. The anticlinal walls of ordinary epidermal cells curved, sometimes nearly in Ω shape. The periclinal walls flat or slightly convex, with papillae or scattered trichomes. Trichome obtuse conical, up to 20 μm long, base diameter up to 13 μm. Stomata mainly distributed at the leaf base and scattered near the leaf edges. Stomata amphicyclocytic, with the guard cells strongly sunken. Subsidiary cells 4-7 in number, cuticularly thickening or papillate.

In lower cuticle, the ordinary epidermal cells usually elongated in form, about 48-69 μm × 13-28 μm in size, with the anticlinal walls straight or U-shaped curved. The periclinal walls convex, often thickened or papillate, sometimes with a central papilla for each, about 8-10 μm in diameter. The stomata cyclocytic, about 21-29 μm × 19-24 μm for each size, irregular in arrangement, with not orientation. The guard cells strongly sunken. Subsidiary cells 4-6 in number for each stoma, strongly papillate, almost covering the stomatal aperture (Gong, 2007).

The present material is different from the Cretaceous *G. adiantoides*, *G. pilifera*, *G. pluripartita* and other known species in the typical amphistomatal cuticles. Due to the fact that the material needs to be further studied, the original author regarded this present specimens as *Ginkgo* sp. (Gong, 2007).

Horizon: Upper Cretaceous, Taipinglinchang Formation.

Coniferales

Parataxodium sp. (Fig. 13, 1)

Leafy shoots. Axis usually about 2 mm wide. Leaves linear in form, 5-15 mm long by 1-2 mm wide for each; alternate and pinnate, streching with a descended base from the axis at an angle of 30°-40°, apical to obtuse in apex. Leaves with a midvein for each, not very clear.

In gross morphology, the leafy shoots of this taxon are similar to *Metasequoia* and *Taxodium*, but different in having relatively thick and some straight axes.

Horizon: Upper Cretaceous, Yong'ancun and Taipinglinchang formations.

Taxodium olrikii (Herr) Brown * (Fig. 13, 2-3)

This taxon was described by Zhang (1984, p. 120, pl. 3, figs. 6-8, 10). The leaf needle-like or linear-lanceolate, single veined, 1.5 cm long by 1 mm wide, acute in apex, base contracted down to axis, arranged in two rows. The specialized leaves of reproductive branches, small, spirally arranged, with a midvein, and the cones about 3 mm in diameter, scaly, probably male cone.

The present specimens newly discovered (Fig. 13, 3) are very similar in gross morphology to the specimens described by Zhang (1984).

Horizon: Upper Cretaceous, Yong'ancun and Taipinglinchang formations.

Metasequoia disticha (Heer) Miki (Fig. 13, 4-8)

Leafy shoots, 3-5 cm long by 1.5-2.5 cm wide, and the branchlet axis about 1-2 mm wide. Leaves opposite at axis usually, with an angle of about 40°-60°, linear, flattened, about 10-15 mm long by 2-3 mm wide, apically obtuse, basally contracted and rounded, with a short stalk. Midvein clear and straight.

The present specimens are characterized by the leaves pinnately opposite at the axis, with a contracted short stalk in leaf base. In gross morphology, the present material is basically consistent with the type material of this species described by Heer (1878, p. 33, pl. 8, fig. 25b; pl. 9, fig. 1; p. 52, pl. 15, figs. 10-12).

Horizon: Upper Cretaceous, Yong'ancun and Taipinglinchang formations.

Sequoia sp. (Fig. 13, 9-11)

Leafy shoots, about 2-5 cm long by 0.7-1.5 cm wide, with branchlet axis about 1.5 mm in width. Leaves linear to elliptic in form, obtuse in apex, 5-15 mm long by 1-3 mm wide for each, helically arranged, decurrent in base and streching from the branchlet axis with an angle of about 45°. Midveins of leaves unclear.

The leaves spirally arranged and decurrent in the base are the main features.

Horizon: Upper Cretaceous, Yong'ancun and Taipinglinchang formations.

Glyptostrobus sp. (Fig. 13, 12-13; Fig. 14, 13)

Leafy branchlets, about 3-11 cm long by 1-1.6 cm wide, with thin axis about 1-1.5 mm wide, and thin leaves. Leaves linear, 4-10 mm long by 1 mm wide, alternate, with a small angle and decurrent to the axis. Midvein not distinct. A cone scattered preserved, fan-shaped, about 5 mm long by 3 mm wide in size.

Horizon: Upper Cretaceous, Yong'ancun and Taipinglinchang formations.

Larix sp. (Fig. 14,1)

Leafy shoots, 7-8 leaves clustered on a short shoot. Leaves linear and needle-like, 8-12 mm long by 0.5-0.7 mm wide for each. Midvein not clear. Short shoot about 3-7 mm long by 4 mm wide. Reproductive parts not preserved.

The present specimens are similar to some living species (e.g. *Larix sibirica* Led.) in gross morphology, but due to no reproductive parts in preservation, it is difficult to make a further comparison.

Horizon: Upper Cretaceous, Taipinglinchang Formation.

Pityospermum minutum Samylina (Fig. 14, 2)

Samara obovate, 7 mm long, tapering upward, obtuse in apex. Wing surface with longitudinal stripes. Seeds born from the wing base, oval, the radius of about 3 mm; wings usually long triangle.

Horizon: Upper Cretaceous, Yong'ancun and Taipinglinchang formations.

Cupressinocladus sveshnikovae Ablaev (Fig. 14, 4-8)

Leafy shoots. The shoot branch axis 3-5 mm wide, branchlets alternate. Leaves scaly, 5-15 mm long by 2-3 mm wide for each, obtuse in apex, spirally arranged on the branchlets.

The present specimens are basically consistent in gross morphology with the type specimens of the species of Russia (Ablaev, 1974).

Horizon: Upper Cretaceous, Yong'ancun and Taipinglinchang formations.

Elatocladus sp. 1 (Fig. 14-9)

Leafy shoot, over 2 cm long by 1.5 cm wide seen; axis thin, about 1 mm wide. Leafy branches broadly linear in form, about 1 cm long by 1.5-2 mm wide for each, with the axis about 0.5 mm wide. Leaves nearly opposite in arrangement, nearly broad subfalcate with some curved upward, about 1-1.5 mm long by 0.5-0.7 mm in the widest base for each, apically acuminate, subspiculate. Midveins unclear.

The gross morphological characters of the present leafy shoot are quite similar to the living heterophyllous leafy shoots of *Taiwania flousiana* Gaussen from Yunnan of China (Sveshnikova, 1963, p. 219, pl. 15, fig. 5) and the leaves of *Taiwania* from the Eocene of southwestern Ukarine reported by Sveshnikova (1963, pl. 15, fig. 12). However, since the present material has no reproductive organs preserved, it is difficult in further classification for the moment, and with no exclusion of a new Cretaceous taxodiaceous taxon for the present material.

Horizon: Taipinglinchang Formation.

Elatocladus sp. 2 (Fig. 14, 10-11)

Branches slender, about 0.8 mm wide. Leaves linear, 7-12 mm long by about 1 mm wide for each, slightly curved upward, and spirally arranged at the axis, with a midvein thick and straight for each.

Horizon: Upper Cretaceous, Yong'ancun and Taipinglinchang formations.

Carpolithus sp. (Fig. 14-12)

Fossil seed, nearly spherical in form, about 7 mm in long diameter. The exotesta thin shelled with a smooth surface. The seed scattered and incompletely preserved, unknown in classification.

Horizon: Upper Cretaceous, Taipinglinchang Formation.

cf. *Podocarpus tsagajanicus* Krassilov

Leaf broadly linear to lanceolate, 3.4 cm long by 3-4 mm wide, apically obtuse, slowly narrowing downwards, with a single vein. Detailed description of this species seen in Zhang (1984).

Horizon: Upper Cretaceous, Taipinglinchang Formation.

Pityophyllum sp.

Leaf linear, 2.5 cm long by 3 mm wide, apically obtuse, with single midvein, and not preserved in its base.

Horizon: Upper Cretaceous, Taipinglinchang Formation.

Angiospermae

Araliaephyllum? sp. (Fig. 15, 1)

Leaf simple, entire, apex notched, base cuneate with a short petiole, leaf veins pulp-pinnate, main veins straight; secondary veins opposite, curved upward to the leaf margin; tertiary veins unclear.

Horizon: Upper Cretaceous, Yong'ancun Formation.

Arthollia tschernyschewii (Konstantov) Golovneva, Sun et Bugdaeva (Fig. 15, 2; Fig. 18, 1-3)

Leaf simple, entire, oval to elliptic, base cuneate or contracted cordate, often with a short stipe, apically acuminate, 5-15 cm long by 3-10 cm wide. Leaf margin finely toothed and slightly concave. Veins pinnate, middle veins straight, secondary veins in 7-8 pairs, alternate. Tertiary venations consisting of widely spaced scalariform anastomoses.

The small size and cordate base leaf shape in the present specimen are distinct from *A. pacifica* and *A. inordinata*. In addition, compared with *A. pacifica*, the present taxon has more bifurcations in secondary veins at the leaf margin (Golovneva et al., 2008).

Horizon: Upper Cretaceous, Taipinglinchang Formation.

Arthollia orientalis (Zhang) Golovneva (Fig. 15, 3)

Leaf simple, entire, broadly ovate, lower part broad, 8-9 cm wide by 7-8 cm long, base cuneate or cordate, apex obtuse. In the palmate pinnate veins, the middle veins straight to the top, the lateral main veins developed; the secondary veins slightly curved, and the tertiary veins forming scalariform anastomoses structure.

This taxon was originally named *Pterosperites orientalis* Zhang (Zhang, 1984), and combined by Golovneva to *Arthollia* in genus (Golovneva et al., 2008).

Horizon: Upper Cretaceous, Taipinglinchang Formation.

Celastrinites kundurensis Golovneva, Sun et Bugdaeva (Fig. 15, 4-8)

Simple leaf, stipitate, elliptic or lanceolate, 3-18 cm long by 1-6 cm wide, apically rounded or apiculate, basally cuneate or truncate, often asymmetrical. The upper part leaf margin small teethed or serrated, lower part entire. The venation pinnate and brochidodromous. The midvein straight, relatively thick, particularly in the lower part, reaching 4 mm wide. 8-11 pairs of secondary veins, alternating or connivent in pairs, straight or slightly arching, subparallel, deviating from the midvein by an angle of 40°-50° and contacting near the margin with the formation of a series of diminishing loops. The tertiary veins formed sca-lariform anastomoses with a polygonal reticulum.

Compared with *C. insignis* from Paleocene North America, the leaves of the latter are larger (up to 20 cm), with dentate margins from base to top, more secondary veins (10-15 pairs), diverging from the midveins by larger angle (50°-80°). *C. septentrionalis* is very similar to the present species, belonging to "elongated leaf", but different in more varia-tions of the leaf morphology, and the leaf base as the widest, and the stratification tends to be newer (Golovneva et al., 2008).

Horizon: Upper Cretaceous, Taipinglinchang Formation.

Platanus densinervis Zhang (Fig. 15, 9)

Macrophyll. Leaf wide and elliptic, nearly entire, more than 5 cm long by about 3 cm wide in the middle and lower parts. Midvein straight, thin, with fine lateral veins branching pin-nately from the midvein with an angle of about 60°-70°. Tertiary veins reticulate but not well defined (Zhang, 1984).

Horizon: Upper Cretaceous, Taipinglinchang Formation.

Platanus sinensis Zhang (Fig. 15, 10-11)

Large leaves trifid, lobes obtuse or obtuse in apex, margin toothed. Palmately three-veined in venation. Middle veins straight, secondary veins in 3-5 pairs, opposite or subopposite, all thinning to the leaf margin. Tertiary veins extending at wide angles, forming a rectangular grid (Zhang, 1984).

Horizon: Upper Cretaceous, Taipinglinchang Formation.

Platanus sp. (Fig. 15, 12)

Leaf simple, oblate, 4.6 cm wide by 3.6 cm high, upper trilobed, lower truncate, base slight-ly decurrent, margin with shallow concave teeth. Palmately three-veined in venation. Mid-dle veins reaching the apex; lateral main veins extending at about 45° from the base and slightly curved to the apex of both lobes; secondary veins pinnate, alternate, tertiary veins indistinct (Zhang, 1984).

Horizon: Upper Cretaceous, Yong'ancun and Taipinglinchang formations.

Dalembia jiayinensis Sun et Golovneva (Fig. 16)

Compound pinnately odd leaves, with 5 leaflets. Lobules elliptic, ovate or subtriangular, ob-tuse in apex, lobules base cuneate, truncate or cordate, slightly asymmetric. Leaf margin entire or lobed. Venation pinnate or palmately pinnate, with lateral veins straight or semi-straight.

Previously this genus was found only in the Cenomanian-Coniacian of Russian Far East (Lebedev & Herman, 1989). The discovery of the genus in Jiayin, China, has expanded the paleogeographic distributions and refreshed the understanding of its stratigraphic range.

Horizon: Upper Cretaceous, Yong'ancun Formation.

Nyssidium arcticum (Heer) Iljinskaja (Fig. 17, 1)

Racemose infructescence, ovoid, 4-8 cm long by 2-3 cm wide; composed of 4-16 spirally arranged follicles, shortly stalked about 1 mm long. Follicles ovoid or elliptic, 5-8 mm long by 4 mm wide, tapering in the middle to both ends, with fine parallel long stripes on the surface.

The present specimen is similar to *N. jiayinense* reported by Feng et al. from Wuyun Formation in Jiayin (Feng et al., 2000), but the latter has a large number of follicles and transverse ridges on the surface of the inner pericarps.

Horizon: Upper Cretaceous, Yong'ancun Formation.

Trochodendroides arctica (Heer) Berry (Fig. 17, 2-4; Fig. 18, 6)

Leaf simple, long ovate or elliptic, with a longer petiole. Leaf apex obtuse, base cuneate, circularly cuneate, or cordate, margin crenate or undulate. Leaf venation palmately three-veined, while some wider leaves palmately five-veined. Middle veins reaching the tip of the leaf. Lateral veins arcing curved, and the tertiary veins pinnately branched, and often forming scalariform reticulum.

The vein characteristics of present specimens are similar to those of *Cercidiphyllum*, but the fruit of *Cercidiphyllum* is usually simple fruit and racemose infructescence, which is quite different from the present taxon.

Horizon: Upper Cretaceous, Yong'ancun and Taipinglinchang formations.

Trochodendroides lanceolata Golovneva, Sun et Bugdaeva (Fig. 17, 5-10; Fig. 18, 5)

Leaf simple, mostly lanceolate in form, incomplete in apex in preservation, entire in margin, and narrower and cuneate in the lower part to the base, asymmetric in form. The venation palmate and brochidodromous.

Horizon: Upper Cretaceous, Taipinglinchang Formation.

Trochodendroides taipinglinchanica Golovneva, Sun et Bugdaeva (Fig. 17, 11-13)

Leaf simple, leathery, rounded to broadly elliptic, 1-7 cm long by 0.8-6 cm wide, short-stalked. Leaf base broadly cuneate, leaf upper part with regular and small crenate of 3-4 mm in diameter. Venation palmate and brochidodromous. Middle veins straight, and 2 lateral veins stretched from the petiole to leaf margin, and forming loops. Tertiary veins more dense, usually forming scalariform reticulum.

The taxon is characterized by small regular crenate at middle-upper parts of leaf, wide wedge at the bottom, and short stalks (Golovneva et al., 2008).

Horizon: Upper Cretaceous, Taipinglinchang Formation.

Trochodendroides microdentatus (Newberry) Krysht. (Fig. 17, 14; Fig. 18, 7)

Leaf simple, broadly ovate, 3-10 cm long by 2.5-5 cm wide, acute in apex, with obtuse teeth, teeth notch obtusely round. Palmately three-veined. Middle primary vein straight, upward thinning. Later primary veins on both sides weak, and often bifurcating for 2-3 times near the leaf margin. The bifurcation veins and the tertiary veins forming polygons. The tertiary veins arranged in an irregular grid.

The apex features of the present specimen are very similar to the type specimen described by Kryshtofovich (1966), and similar to those of this taxon from Zeya-Bureya basin reported by Krassilov (1976) and Jiayin (Zhang, 1984).

Horizon: Upper Cretaceous, Yong'ancun and Taipinglinchang formations.

Menispermites sp.

Leaf simple, peltate or reniform, about 4-6 cm long by 5-7 cm wide, apex obtuse, base contracted, broadly cordate, petiolate. Upper leaf margin undulate and crenate, base nearly entire. Palmately 5-6 basal veins, middle veins straight, lateral veins curved, and bifurcating for 2-3 times to leaf edge. Tertiary veins weak, interwoven into irregular fine meshes.

Horizon: Upper Cretaceous, Yong'ancun Formation.

Viburnum cf. *contortum* Lesquereux* (Fig. 18, 8)

Leaf simple, ovate, unlobed, 9 cm long by 7.5 cm wide, apex obtuse, base broadly cuneate, upper margin undulate, lower entire. Venation pinnate, the middle veins straight to apex, secondary veins in 7-8 pairs, subalternate. Tertiary veins forming rectangular meshes (Zhang 1984).

Horizon: Upper Cretaceous, Taipinglinchang Formation.

Viburnophyllum sp. (Fig. 18, 9)

Leaf simple, elliptic, 4 cm long by 3 cm wide, base broadly cuneate, margin serrulate. Venation pinnate. Middle veins straight, secondary veins in 4-5 pairs, subalternate. Tertiary veins at the leaf margin, clear and arcing curved to the edge, forming irregular meshes.

Horizon: Upper Cretaceous, Taipinglinchang Formation.

Nordenskioideia cf. *borealis* Heer

Fossil fruit, round, about 7 mm in diameter, with about 6 lobes tightly bound, giving the fruit a colyliform shape (Zhang, 1984).

Horizon: Upper Cretaceous, Taipinglinchang Formation.

Aquatic angiosperms

Cobbania corrugata (Lesq.) Stockey, Ruthwell et Johnson (Fig. 19)

Aquatic herb with rosette compound leaves. Leaflet suborbicular or elliptic, slightly undulate or dentate throughout, 1-9 cm long by 1-6 cm wide. The leaf adaxial surface curved and thin, and the leaf margin with regular subcircular veins. The primary veins, secondary veins, and tertiary veins of the leaf interwoven into a network, forming an irregular polygonal fine mesh. In the leaf abaxial surface, the primary veins up to 7.6 cm long, the secondary veins weak and interweaved with the tertiary veins, forming irregular meshes. Petiole thick, about 5-9 mm in diameter. No root and reproductive parts preserved.

The morphological features of the present specimen leaves are basically consistent with those of the typical specimens described by Stockey et al. (2007).

Horizon: Upper Cretaceous, Yong'ancun and Taipinglinchang formations.

Nelumbo jiayinensis Liang, Sun et Yang (Fig. 20)

Aquatic herbs plants. Leaf orbicular peltate, entire or slightly undulate in the margin. The petiole located in the center of the blade, with 20-25 primary veins, extending radially from the center of the blade to the leaf margin, and bifurcating 1-2 times near the margin. Secondary veins curved and interlaced into a network. Tertiary veins weaken and weaved into fine meshes. The morphology of leaf epidermis cells observed by superdepth microscopy in the upper epidermis. The ordinary cells regularly polygonal in form, about 20-50 μm in size for each, with the anticlinal wall straight. In the lower epidermis, the cells similar to that in the upper epidermis, about 40 μm × 30 μm in size for each, and no stomata observed.

Horizon: Upper Cretaceous, Yong'ancun Formation mostly, and a few from Taipinglinchang Formation.

Quereuxia angulata (Newb.) Kryshtofovich (Fig. 21)

Aquatic herb plants, microphyll. Leaves heteromorphic, including submerged and floating. Floating leaflets broadly ovate to narrowly obovate, margin often toothed, petiole slender, veins pinnate. Submerged leaflets linear, radiating, false bifurcating for many times. Samylina (1988) proposed a restoration of the ecological characteristics of the genus (Fig. 21, 11). The leaf epidermis cells hyperstomatic in type. The normal epidermal cells on the inner surface of the upper epidermis mostly rectangular (about 50-70 μm × 30 μm) or regularly polygonal (4-6 sides), with slightly straight anticlinal walls and a few stomata. Stomata elliptic and oriented with 4-6 subsidiary cells. The ordinary epidermal cells on the outer surface mostly longitudinally folded, with stomata oriented in arrangement. Folded longitudinal cuticularization seen outside the guard cells. The ordinary epidermal cells in the lower epidermis mostly polygonal (4-6 sides) in form, with no stomata.

Horizon: Upper Cretaceous, Yong'ancun and Taipinglinchang formations.

2.1.2 Sporopollen fossils

With the discovery of a large number of plant megafossils, abundant sporopollen fossils have been found in the flora, which have played an important role for studying the composition and age in detail of the Late Cretaceous flora in Jiayin, Heilongjiang Province.

The Late Cretaceous sporopollen fossils in Jiayin can be divided into five assemblages: *Kuprianipollis santaloides*-*Duplosporis borealis* Assemblage (represented by Yong'ancun Formation, aged in Santonian), *Aquilapollenites conatus*-*Podocarpidites multesimus* Assemblage (represented by Taipinglinchang Formation, aged in Campanian), *Aquilapollenites amygdaloides*-*Gnetaceaepollenites evidens* Assemblage (represented by lower Yuliangzi Formation, aged in early Maastrichtian), *Wodehouseia aspera*-*Parviprojectus amurensis* Assemblage (represented by upper Yuliangzi Formation, aged in middle Maastrichtian), and *Aquilapollenites conatus*-*Pseudoaquilapollenites striatus* Assemblage (represented by Furao Formation, aged in late Maastrichtian) (Figs. 22-24).

2.2 Floristic characteristics

The middle-late Late Cretaceous flora of Jiayin is mainly composed of mosses, ferns, gink-gos, conifers, and angiosperms. Among them, bryophytes and ferns usually grow in the lowlands of rivers and lakes, showing a humid environment; but the appearance of *Gleichenites* seems to reflect the relatively humid and hot climate. A large number of *Ginkgo* also reflect the warm and humid environment. For conifers, some abundant Taxodiaceae plants, such as *Metasequoia*, *Sequoia*, *Taxodium*, and *Glyptostrobus*, mainly grow in a warm temperate environment. Among the angiosperms, *Platanus*, *Trochodendron*, *Celastrinites*, and *Viburnum* are also growing in warm temperate zone. Most of these plants have broad and entire leaves. Their deciduous leaf shows subtropical and seasonal climates. The appearance of a large number of aquatic angiosperms, such as *Quereuxia*, *Cobbania*, and *Nelumbo*, seems to indicate that there were sufficient water supply sources at that time.

According to the characteristics of some angiosperms in the flora, the leaves of Late Cretaceous angiosperms in Jiayin are usually smaller in size. For example, the longest leaf length of *Platanus* is only 6-8 cm long, and that of *Trochodendroides* is generally less than 4-5 cm, which seems to indicate the characteristics of a small leaf type. Except some taxa with entire leaves (e.g. *Platanus*) and coriaceous leaves (e.g. *Araliophyllum*) which may indicate hotter climate, more angiosperms have thin and some toothed leaves, indicating that the climate may be relatively mild and humid, and may relatively cool in season. In Taipinglinchang assemblage, a big amount of *Classopollis* pollen indicate that the flora may have experienced seasonal drought in Campanian. Markevich explained that the appearance of thermophilic taxa in the Late Cretaceous flora may be affected by the warm, humid and hot marine climate in the Jiayin area during the middle-late Late Cretaceous (Markevich et al., 2005, 2006).

Generally, the flora of Jiayin in the middle-late Late Cretaceous may belong to a warm temperate flora, with seasonal changes, similar to the low mountain warm temperate forest in the Yangtze River Basin of China nowadays. The early assemblage (Yong'ancun assemblage, Santonian) may be more humid and hotter, while the late assemblage (Taipinglinchang assemblage, Campanian) is relatively mild (Fig. 25).

2.3 Geological age

According to the comprehensive study of mega-plant and sporopollen fossils, the Late Creta-ceous flora in Jiayin belongs to the middle-late Late Cretaceous (Santonian to Campanian) in age. The age can be evidenced by (1) the plant fossils, particularly the sporopollen fos-sils; (2) the correlation of the flora-bearing strata in Jiayin with marine or marine alternat-ed non-marine beds in Far Eastern Russia yielding the contemporary floras; and (3) some isotopic datings related.

(1) Geological age indicated by plant fossils, particularly by sporopollen

Most of the plants from the Upper Cretaceous in Jiayin are common elements of the Late Cretaceous floras, including some important elements limited to the Late Cretaceous age, such as angiosperms *Dalembia*, *Arthollia*, *Celastrinites*, *Cobbania*, *Araliaephyllum*, *Trocho-dendroides taipinglinchanica*, etc., and gymnosperms (e.g. *Parataxodium*, *Ginkgo pilifera*, etc.), which have been only found in the Upper Cretaceous in Northeast Asia. *Platanus* flourished since the Cenomanian and were widely distributed in the Upper Cretaceous in Northeast Asia and North America. *Gleichenites* flourished in Mesozoic but were almost ex-tinct by the end of the Late Cretaceous. As for the other plant fossils, such as *Ginkgo adian-toides*, *Metasequoia disticha*, *Sequoia* sp., *Glyptostrobus* sp., *Trochodendroides arctica*, *Quereuxia angulata*, etc., they are also common fossils of Late Cretaceous floras, although they can continue to the Paleocene, even younger, in geological range.

The strong evidence for the age is given by the sporopollen fossils from the Late Creta-ceous flora in Jiayin, in which most of them show the Late Cretaceous age, particularly some important sporopollen taxa are limited to the Late Cretaceous age, such as *Aquilapol-lenites amplus*, *A. conatus*, *A. reductus*, *A. rigidus*, *A. rombicus*, *A. stelkii*, *A. striatus*, *Intergri-corpus bellum*, *Marsypiletes cretacea*, *Parviprojectus amurensis*, *Proteacidites mollis*, *Pseu-dointegricorpus clarireticulatus*, *Wodehouseia aspera*, *W. gracilis*, etc. (Figs. 26, 27). Be-sides, some index sporopollen fossils have been only found in the Maastrichtian, such as *Aq-uilapollenites funkhouseri*, *A. reductus*, *Pseudoaquilapollenites striatus*, *Pseudointegricorpus clariretuculatus*, *Marsypiletes cretacea*, *Zlivisporites novomexicanus*, *Rugulatisporites quin-tus*, *Wodehouseia stanley*, etc. Among them, *Aquilapollenites reductus*, *Marsypiletes cretacea*, etc. are the latest extinct in the Maastrichtian (Fig. 27).

(2) Evidenced by correlation of the flora - bearing strata in Jiayin with marine or marine alternated non-marine beds yielding the contemporary floras in Far Eastern Russia

The strata yielding the contemporary flora are represented by the Kundur Formation and up-per Zavitin Formation in the Zeya-Bureya basin of Russian Far East. The fossil assemblage of the Kundur Formation is very similar to that of Taipinglinchang Formation of Jiayin, and contains at least 20 common taxa, such as *Ginkgo pilifera*, *Sequoia* sp., *Metasequoia* sp, *Cu-pressinocladus* sp., *Trochodendroides lanceolata*, *T. taipinglinchanica*, *Arthollia orientalis*, *Celastrinites kundurensis*, *Quereuxia angulata*, *Cobbania corrugate*, etc. The Kundur Forma-tion is documented as the Campanian age (Golovneva et al., 2008). Particularly, among the above assemblage *Ginkgo pilifera* did not exceed the Campanian stage (Samylina, 1967; Golovneva, 2005). In addition, *Celastrinites* was first found in the Santonian, but most characteristics in Maastrichtian; *C. septentrionalis* (Krysht.) Gol. flourished in the Kakanaut Formation of middle Maastrichtian in the Koryak Upland, which also produces dinosaurs (Golovneva, 1994). *Cobbania corrugate* is a common taxon in the Campanian-Maastrichtian in North America (Bell, 1949; Johnson, 2002), and also found in the lower Campanian in the Khatanga River area of northern Siberia (Abramova, 1983; Golovneva, 2005), in which the index-bivalves *Inocermus patotensiformes* were also found in this hori-zon (Golovneva et al., 2008). Considering that the dinosaur-bearing strata Yuliangzi For-mation and Lower Tsagajan Formation are covering Taipinglinchang Formation and Kundur Formation, respectively, their age should be the early to middle Maastrichtian. Therefore, the age of the Taipinglinchang floral assemblage and the late Kundur floral assemblage seems to be the Campanian, and the sporopollen research results also support this age (Markevich et al., 2005, 2006; Sun et al., 2014). In Northeast Asia, the Zonk'erian flora of Sakhalin Island (the early Campanian; Krassilov, 1979), the Barikovian flora (the early Santonian-Campanian; Herman & Lebedev, 1991), the Mutino flora (the early Campanian), and Sym flora (the Campanian; Abramova, 1983; Golovneva, 2005) are similar to the Taipinglinchang and Kundur flora in age.

(3) Isotopic dating evidence related

Although there are no direct isotopic data for the Late Cretaceous flora in Jiayin, the relevant isotopic dating results and stratigraphical correlation could be used to refer to the age consideration of the Late Cretaceous flora in Jiayin. Among them, the most important data for correlation come from the Songliao Basin in the western Heilongjiang Province (Wan et al., 2013; Xi et al., 2019).

According to Wan et al. (2013) and Xi et al. (2018, 2019), the Yong'ancun and Taipinglinchang formations of Jiayin can be correlated with the Yaojia and Nenjiang formations of the Songliao basin, respectively. The two formations of the Songliao basin are aged in the Santonian-Campanian, in general, within the scope of the isotopic dating as 86-80 Ma. The correlation could be referred to as the Yong'ancun and Taipinglinchang Formations and their floral assemblages are coincident with the middle-late Late Cretaceous (i.e. the Santonian-Campanian) in age. Moreover, not long before, the authors found the fossil fish, *Sungarichthys longicephalus* Takai, from the Taipinglinchang Formation in the Target Range No. 4 of Jiayin (Figs. 5-5, 6). This fish taxon was found from the Fulongquan Formation (equals to Nenjiang Formation) of Jilin, which provided an important evidence for the correlation of the Taipinglinchang Formation of Jiayin with the Nenjiang Formation of the Songliao basin.

2.4 Recent advance on research

During 2015-2018, with the support of the MOST and CGS for the projects "study of Cretaceous-Paleogene boundary in China", the authors have achieved a lot in the study of the Late Cretaceous flora in Jiayin, which mainly includes: (1) discovery of Late Cretaceous angiosperms *Dalembia*; (2) discovery of the aquatic angiosperms *Nelumbo* in the Upper Cretaceous; (3) study on the leaf epidermal structure of the aquatic angiosperms *Quereuxia*.

(1) The discovery of *Dalembia* in Jiayin

In the past, *Dalembia* was only found in the Cenomanian-Coniacian in the Far East of Russia (Herman & Lebedev, 1991). The discovery of *Dalembia* in Jiayin has expanded the understanding of the age and palaeogeographic distribution of the genus for the first time: the age of this genus can reach the Santonian in the Late Cretaceous, and the southernmost point of the fossil distribution can reach Jiayin of Heilongjiang, China. This achievement also provides new evidence for the Late Cretaceous stratigraphical correlation between

China and the Far Eastern Russia (Sun et al., 2016) (Figs. 16, 28).

(2) The discovery of aquatic angiosperm *Nelumbo* in Jiayin

From 2017 to 2018, the author first found the aquatic angiosperms *Nelumbo* in the Late Cretaceous Yong'ancun Formation in Jiayin, with naming a new species *Nelumbo jiayinensis*, published in the journal *Cretaceous Research* (Liang et al., 2018a) (Fig. 29).

Nelumbo (the lotus) is one of the ten most famous flowers in China. The earliest fossil *Nelumbo* in the world has been recorded in the Lower Cretaceous (Albian) in the United States and Portugal. In China, the earliest records of *Nelumbo* fossils were found in the Eocene in the Hainan Island, Yilan of Heilongjiang, and Fushun of Liaoning. Thus, the discovery of *Nelumbo* in Jiayin has advanced the record of this taxon in China for at least 30 million years. Meanwhile, As *Nelumbo* has mainly grown under subtropical or warm temperate climates, the new discovery is of great significance for the restoration of the Late Cretaceous paleogeography and paleoclimate in Jiayin and its neighboring Russian areas, as well as beneficial for the study on the origin and evolution of *Nelumbo*, even for the phytogenetic study of angiosperms in China and the world.

(3) The discovery of the epidermal structure of *Quereuxia* in Jiayin

Quereuxia is an aquatic angiosperm, first discovered in the Cretaceous and Paleocene in North America. For a long time, this genus was regarded as the taxon *Trapa*. However, there are obvious differences between *Quereuxia* and *Trapa*. The floating leaves of *Quereuxia* are rosette shaped compound leaves, while the leaves of *Trapa* are single, and the fruits of the two taxa are significantly different from each other (Liang et al., 2018a). Although the research on *Quereuxia* has been more than one hundred years, there is a lack of study of the leaf epidermal structure. In recent years, by using the VHX-5000 3D microscope, the authors have found the epidermal structure of *Quereuxia* in Jiayin, which would be significant for the study of the taxonomy and evolution of this genus in the Cretaceous, and of the paleoclimate related (Figs. 30, 31).

Chapter 3

Late Cretaceous climates and geography in Jiayin

The Cretaceous (ca. 144-66 Ma) is one of the most active periods of the earth's tectonic movement and evolution and is also one of the most important periods of paleoclimate and biological evolution. Due to global sea-level rise, large areas of land were covered by a warm shallow sea, and the global temperature was higher in the Cretaceous, showing the "greenhouse" effect. Especially in the Middle Cretaceous, the rapid expansion of the mid-ocean ridge led to sea level rise, making the earth even hotter. The global temperature could be 3-10 times as high as that today (Hu, 2004). In the mid-Cretaceous, the slow flow in the lower layers of the ocean resulted in the anoxic environment of the ocean, which is called "ocean anoxic events" (OAE); many black shale formations (e.g. in the North Sea) were formed in this period, and these shale layers are important sources of oil and gas.

Due to the warm climate in the Cretaceous period, especially the humid climate in some coastal areas (e.g. the East Asian Pacific region, including the Far East of Russia and the eastern Heilongjiang of China), favorable conditions were created for the development and prosperity of the vegetation. The great development of angiosperms also provided more abundant food sources for the living and evolution of terrestrial animals such as dinosaurs. In the Late Cretaceous, the global herbivorous dinosaurs represented by Hadrosaurid and

Ceratopsia reached the apex of their evolution, and so did the carnivorous dinosaurs, such as *Tyrannosaurus rex*. In China, a large number of dinosaurs migrated thousands of miles and gathered in the northeastern Heilongjiang region including Jiayin in the middle-late Late Cretaceous, indicating that the climate and ecological environment here were most conducive for living and development of terrestrial creatures such as dinosaurs (Fig. 32).

The Late Cretaceous experienced great changes in the global climate. According to the study of tectonic geology, fossils, and paleo-CO_2, although the global temperature had a sudden rise in the early-middle Late Cretaceous, the "Cenomanian-Turonian thermal maximum" occurred, and the paleo-CO_2 level was relatively high. However, in the middle-late Late Cretaceous (Santonian-Campanian), the global temperature decreased, and gradually down until the end of the Late Cretaceous (Wang et al., 2014; Fig. 32). A large number of studies have shown that in the Late Cretaceous, with the opening of the South Atlantic Ocean, frequent volcanic activities, and large-scale sea regression, the terrestrial ecosystem changed greatly, resulting in the extinction of dinosaurs and a huge number of other bios, and great changes in the terrestrial plant kingdom. This is what is known as "the fifth mass extinction and recovery in the evolution history since Phanerozoic on earth", i.e., the Cretaceous-Paleogene (K-Pg) biological evolutionary event (Sun et al., 2014).

3.1 Late Cretaceous climates

3.1.1 Reflection of plant fossils

As mentioned in Chapter 2 of this book, 43 species of 34 genera fossil plants have been found in the middle-late Late Cretaceous flora in Jiayin, including bryophytes (ca. 2%), ferns (ca. 11%), ginkgos (ca. 7%), conifers (ca. 27%) and, angiosperms (ca. 50%) which became the dominant group in the flora. The above composition represents the outlook of the vegetation in Jiayin in the middle-late Late Cretaceous (Santonian-Campanian).

As for the reflection of plants on climate and environment, the bryophytes and ferns mostly live in humid wetland or swamp areas with sufficient groundwater, which usually reflects a warm and humid climate. However, in the early assemblage (Yong'ancun assemblage) of the flora in Jiayin, the occurrence of *Gleichenites* seems to reflect that the climate was relatively hotter at least in the Santonian.

Regarding *Ginkgo* of the Late Cretaceous flora in Jiayin, although the genus has only 3 species, the number of the genus is large, indicating that it was flourishing in growth during the time. Ecologically, the natural communities of the living *Ginkgo* (i.e. *G. biloba* L.) are mostly distributed and prosperous in the warm temperate environment, near 30° N in southern-central China, such as the mountain foothill regions of the Yangtze River reaches, where they have a warm and humid climate with an average annual temperature of 9-18 ℃ and average annual precipitation of 600-1500 mm (He et al., 1997). The distributions of *Ginkgo biloba* in China are mostly from Shenyang to northern Fujian, where the climates are generally warm and humid. Therefore, the lot of fossil *Ginkgo* in occurrence seems to tell that the climate in the Jiayin area was warm and humid during the middle-late Late Cretaceous.

According to the conifers, the present Taxodiaceae, including *Glyptostrobus*, *Metasequoia*, *Taxodium*, and *Sequoia*, usually grow in a warm temperate climate. The natural community of *Metasequoia - Glyptostrobus* in China is mainly distributed in the Yangtze River reaches with a warm temperate climate, and the precipitation is relatively high. *Sequoia* is an evergreen plant and grows in a warm and humid climate. Based on a large number of Taxodiaceae plants in the Jiayin flora, it seems that the temperature and humidity in the middle-late Late Cretaceous were higher than those in modern times, and there was a warm temperate or warm temperate and temperate mixed climate.

As for the angiosperm, there are a large number of *Platanus*, *Araliaphyllum*, *Trochodendroides*, *Celastrinites*, and *Dalembia* in the Late Cretaceous flora in Jiayin. The living *Platanus* grows in a warm and humid environment, which can be taken as an example of in China. *Platanus* is distributed in the region in the north from the Beijing-Dalian line to the south in the Zhejiang - Fujian border area. Besides, the abundance of aquatic angiosperms in the Jiayin Late Cretaceous flora, such as the large number of *Quereuxia*, *Cobbania*, and *Nelumbo*, reflects not only the abundance of water but also the warm climate in this area.

It should be mentioned that according to the angiosperms with small leaves, few leathery leaves and mostly deciduous plants, and a certain number of pineaceae plants (e.g. *Larix*) in the Late Cretaceous flora of Jiayin, it seems to reflect the characteristics of seasonal change (including seasonal cooling), and the climate may be mainly warm temperate joined by temperate climate seasonally.

It is worth mentioning that the fossil sporopollen in the Jiayin flora have provided more implications on the analysis of the paleoclimatic characteristics reflected by the flora. There are a large number of thermophile and hygrophilic elements of the Late Cretaceous palynoflora in the Jiayin flora (Markevich et al., 2005, 2006), which further indicate the climate in the middle-late Late Cretaceous in Jiayin was dominated by the warm temperate climate. However, based on the occurrence of a certain number of Gnetalean pollen fossils from the Taipinglinchang assemblage, it seems to imply that in the Campanian stage in the Jiayin area, the climate might be slightly dry or experienced seasonal drought, and the temperature may be milder than the Santonian time.

In sum, based on the evidence of the fossils plants, the climate of the middle-late Late Cretaceous in Jiayin and its neighboring areas may be mainly warm temperate with seasonal changes, and the vegetation may be similar to the warm temperate forest in the low mountain areas of the Yangtze River reaches in China today.

3.1.2 The quantitative study of paleoclimate

Fossil plants can provide useful help for the study of paleoclimate. For study of paleoclimate, the effective methods have been by using leaf analysis of fossil plants, including the LMA (Leaf Margin Analysis), CLAMP (Climate-Leaf Multivariate Program) (Wolf, 1995; Wolf & Spicer, 1999), and CA (Co-existence Approach) (Mosbrugger & Utescher, 1997) based on the Nearest Living Relative Method (NLR) (Mosbrugger, 1999). In recent years, some important progress in the study of paleoclimate has been expressed in using the paleo-CO_2 quantitative study. Based on the study of the impact of global carbon cycle changes (including geological periods) on the earth's ecosystem, it is found that the higher the CO_2 content in the atmosphere, the higher the land temperature. Since stomata are tools used to control gas exchange (including CO_2 for photosynthesis) between plants and the outside world, the concentration of CO_2 can be calculated by the stomatal indices (SI) and stomatal density (SD), gained from the cuticular study of the fossil leaves. Results showed that SD and SI of the fossil plants were inversely proportional to CO_2 concentration. Therefore, it has become one of the most effective methods to study the change of CO_2 concentration in the geological history through fossil cuticles, and particularly, the fossil *Ginkgo* is considered as the main taxon for reconstructing paleoenvironment, and its research results have been more reliable.

In recent years, based on a large number of *Ginkgo* fossils collected from the Yong'ancun and Taipinglinchang formations in Jiayin, Quan et al. carried out a high-resolution analysis of the paleo-CO_2 level. A systematic study on the pCO_2 concentration of 77 fossil leaves of *Ginkgo adiantoides* collected from 11 layers (about 5 leaves per layer) of the Taipinglinchang Formation is reported in this paper (Quan et al., 2009; Table 6). According to the sequence analysis of stomatal indices (SI) of leaf cuticles, the pCO_2 of Campanian decreased gradually in the range of 0.055%-0.059% (550-590 ppm). This result is closer to the Geocarb-II model. In the analysis, it was found that there was an obvious short-term fluctuation of CO_2 (SCDF) in the late Campanian stage. The pCO_2 reached 0.069% (690 ppm), but then returned to the level of 0.059% (590 ppm) (Quan et al., 2009). This study is a useful attempt to quantitatively recover the paleoatmospheric CO_2 concentration in the middle and late Late Cretaceous. The above results show that the pCO_2 from Santonian to Campanian in Jiayin area is 1.4-2.5 times higher than that in the modern atmosphere (Fig. 33). Accordingly, the temperature in the middle-late Late Cretaceous in Jiayin is much higher than that in the present time. This result provides strong evidence for the qualitative analy-

Table 6   Stomatal measurement and inferred paleo-CO_2 based on *G. adiantoides* from Cretaceous Taipinglinchang Fm. of Jiayin (after Quan et al., 2009, simplified)

Bed	Leaf no.	SI(mean)	Standard deviation of SI	pCO_2(RF)	RCO_2
11	10	7.01	0.14	558.53	1.62
10	12	6.98	0.27	569.43	1.62
9	7	6.97	0.43	571.03	1.63
8	8	6.92	0.34	588.02	1.64
7	11	6.91	0.27	592.02	1.64
6	8	6.70	0.23	692.02	1.69
5	3	6.96	0.49	575.98	1.63
4	5	6.98	0.31	567.38	1.62
3	7	6.94	0.31	683.55	1.63
2	5	6.93	0.50	584.90	1.63
1	1	6.83	0.49	624.70	1.66

sis of the paleoclimate in Jiayin and its adjacent areas in the middle-late Late Cretaceous.

3.2 Late Cretaceous geography and environment

The study of the Late Cretaceous flora and its bearing strata in the Jiayin area provides important reference for studying the paleogeographic environment of Jiayin and its neighboring area in the middle-late Late Cretaceous.

The sedimentary facies in Jiayin and its neighboring areas show that the areas experienced the evolutionary process of fluvial lacustrine facies (represented by Yong'ancun Formation; Fig. 34)-lacustrine facies (represented by Taipinglinchang Formation)-fluvial facies (represented by Yuliangzi Formation)-and lacustrine facies (represented by Furao Formation), in the middle-late Late Cretaceous. The occurrence of glutenite in the Yong'ancun Formation reflects the development of fluvial facies in the Santonian stage. Certainly, the existence of a large number of mudstone and siltstone also reflects that the lacustrine environment may have begun to take shape at that time. During this time, there were rivers and lakes in Jiayin and its adjacent areas, surrounded by low mountains, and the slopes might be covered with conifers (e.g. *Cupressinocladus*, *Metasequoia*, and *Sequoia*, etc.), ginkgos (e.g. *Ginkgo pilifera*, *G. adiantoides*, etc.) and angiosperms (e.g. Platanoid, etc.). In 2002, the research team discovered a dinosaur track fossil *Jiayinosauropsis* in the coarse-grained sandstone of the Yong'ancun Formation (Dong et al., 2003), which provided evidence that the deposition of the Yong'ancun Formation might reflect fluvial facies, because dinosaurs and other large reptiles usually lived on the beaches of rivers and lakes not far from the forest (Sun, 2019*). The fine clastic sediments (mudstone and siltstone, etc.) represented by the Taipinglinchang Formation reflect that the lacustrine facies in Campanian was widely distributed. The discovery of some fossil invertebrates, such as conchostraca (e.g. *Halysestheria yui* (Chang) = *Halysestheria qinggangensis* Zhang et Chen), ostracods (e.g. *Cypridea perelegana* Ten., *Rhinocypris* cf. *prinaeva* Li, etc.), and fish (e.g. *Sungarichthys longicephalus* Takai) in the Taipinglinchang Formation provides further evidence for the fluvial and lacustrine environments at that time. It can be inferred that the Jiayin area was

---

* Sun G. Mesozoic plants. Text-book for graduates. 2019. Shenyang Normal University.

once in a relatively warm and humid environment, the climate may be warm to moderately temperate, and the precipitation may be higher. The paleogeographic environment may be a shallow lake, lakeshore, sandbank, forested wetland, and piedmont slope land, and the vegetation in the Jiayin area also formed different fossil plant communities with different regions (Sun et al., 2014).

According to the paleogeographic environment reflected by the fossil plants, the floral assemblages of the Yong'ancun and Taipinglinchang formations indicate that the flora at that time may have lived under a warm temperate climate with seasonal changes, in a low mountainous area and surrounded by rivers and lakes. The appearance of deciduous angiosperms (e.g. *Platanus*, *Araliaphyllum*, *Trochodendroides*, *Celastrinites*, and *Dalembia*, etc.) and a large number of warm temperate conifers seem to reflect mountains and slopes. The discovery of aquatic angiosperms, such as *Quereuxia*, *Cobbania*, and *Nelumbo*, indicates that the Jiayin area was once a warm and humid environment with abundant water resources and dense rivers and lakes in the middle-late Late Cretaceous. During the Santonian-Campanian, the Jiayin area may be located in the inland, but not far from the sea, which obtained relatively warm and humid conditions. The association of a large number of herbivorous dinosaurs, represented by hadrosaurid, also indicates that the paleogeography and paleoclimate were warm, humid and well supplied with water in the Jiayin area (Fig. 35).

Besides, a large number of the Late Cretaceous ginkgos (e.g. *Ginkgo adiantoides*, *G. pilifera*, etc.) also have important paleoecological implications. According to the ecological environment, the natural communities of *Ginkgo biloba* are mostly distributed in the warm temperate environment, near 30°N. The most prosperous areas, such as the deep mountain foothills of the Yangtze River reaches, have a warm and humid climate with an average annual temperature of 9-18 ℃ and average annual precipitation of 600-1500 mm (He et al., 1997). In particular, during the Late Cretaceous, Jiayin was located in the area, north of 40°N, it is indicated that the global climate might be warmer at that time, and the surface temperature was much higher than that of modern times. Moreover, Jiayin and the eastern Heilongjiang region were probably affected by the warm current of the paleo-Pacific Ocean in the East, which made the temperature and humidity of the Jiayin area much higher than that of today.

However, on the other hand, the angiosperms in the Late Cretaceous flora of Jiayin were usually with microphylls and unentire leaves, which shows that the climate of Jiayin in the middle-late Late Cretaceous may be dominated by warm temperate climate, but it also has a relatively mild component, and may even have some seasonal cooling. The abundance of conifers reflects that the terrain here may be relatively higher (of course, it may not be far away from the sea), and the flora may mainly grow in the slopeing land and the river lake lowland near the valley bottom (Markovich et al., 2005).

Moreover, based on fossils, the aquatic animals, such as conchostraca, ostracods, bivalves, and fishes were flourishing in Jiayin and its adjacent areas in the middle-late Late Cretaceous. When Russian paleontologist Riabinin (1930) published *Mandschurosaurus*, he reported the discovery of fossil turtle *Aspederetes planicostus* associated with dinosaurs, reflecting the widespread existence of lakes and marshes at that time (Sun et al., 2014). As for the dinosaurs found in Jiayin and its neighboring area, more than 95% of the dinosaurs in Jiayin are herbivorous Hadrosauridae. The warm and humid environment and the prosperity of the vegetation were the guarantees for their survival and development. A large number of studies have shown that dinosaurs mainly lived in a warm and humid climate, not far from the forest in the beaches of the rivers and lakes, which provides more evidence for the restoration of the middle-late Late Cretaceous paleogeographic environment in the Jiayin area (Sun et al., 2014) (Fig. 35).

Chapter **4**

Late Cretaceous flora with the K-Pg boundary in Jiayin

Dinosaurs lived mainly in the forest areas of lakeshore plain or the open area with luxuriant plants. In the middle-late Late Cretaceous, because of the flourishing vegetation and warm and humid climate, many dinosaurs, especially a large number of herbivorous dinosaurs, migrated to Jiayin, eastern Heilongjiang region from Northwest China, North China, and Southeast China. In Jiayin, more than 95% of the dinosaurs are herbivorous represented by Hadrosaurid, and at least 10 taxa have been found (Dong et al., 2003; Wu et al., 2010; Godfroit et al., 2011). Besides, there are some carnivorous dinosaurs represented by Tyrannosaurid (e.g. *Tyrannosaurus*, etc.). The Hadrosaurid mainly include (Fig. 36):

Hadrosaurinae

① *Mandschurosaurus amurensis* (Riab.)Riabinin, 1930

② *Mandschurosaurus magnus* Zhao, 1995 (MS)

③ *Jiayinosauropis johnsoni* Dong et al., 2003

④ *Kerberosaurus manakini* Bolotsky et Godefroit, 2004

⑤ *Wulagasaurus dongi* Godefroit et al., 2008

⑥ *Kundurasaurus* Bolotsky et Godefroit, 2014

Lambeosaurinae

⑦ *Charonosaurus jiayinensis* Godefroit et al., 2000

⑧ *Sahaliyania elunchunorum* Godefroit et al., 2008

⑨ *Amurosaurus riabinini* Bolotsky et Kurzanov, 1991

⑩ *Olorotitan arharensis* Godefroit et al., 2003

The herbivorous dinosaurs include all Ornithischia and most of Saurischia dinosaurs, which are usually large in size. For example, the largest known dinosaurs in China, *Ruyangsaurus* from Henan, can reach 38 m long; the length of *Xinjiangtitan* from Shanshan of Xinjiang is nearly 30 m (Dong, 2009; Wu et al., 2013). Such a large body need to have enough food supply. By analysis, the huge dinosaur *Diplodocus* weighs about 30 tons, which is over 10 times as heavy as living elephants, and might consume about 1 - 1.5 tons of plants per day. Such a large amount of food consumption requires a very prosperous plant kingdom to satisfy their food supply and survival needs (Sun, 2019*). Thus, which plants were the "dining table food" for the Jiayin's herbivorous dinosaurs at that time?

4.1 Dining table for herbivory dinosaurs

In terms of morphological and functional characteristics, herbivorous dinosaurs in Jiayin were usually large, and most of them had long necks that made it easy for them to eat from high trees. For example, the dinosaurs *Mandschurosaurus* and *Charonosaurus* found in Jiayin could be 11 m long, with the neck of at least 4 - 5 m in length. *Olorotitan* was about 8 m long, and its neck was composed of 18 neck bones. Their forging height might reach 6 - 10 m, and we can reasonably infer that the height of plants they ate was at least 6-10 m or higher. The teeth of herbivorous dinosaurs were spoon or nail in shape, which would be convenient for cutting off leaves or leafy shoots, even piercing the cone shells (Fig. 38, 7). These teeth shapes indirectly indicate that there were cone-bearing plants in the vegetation at that time, such as the conifers and cycads (Fig. 38, 2, 4, 8, 10). In particular, the dentaries of *Mandschurosaurus* and *Olorotitanwere* make it easy to grind (Fig. 38, 5, 6). Of course, some plant seed coats (e.g. ginkgos and cycads) or megasporophylls (e.g. cycads), and a large number of delicate ferns and angiosperms, especially the leaves of aquatic flowering plants (e.g. *Quereuxia*, *Cobbania*, and *Nelumbo*) are relatively soft and easier for dinosaurs to chew and digest (Fig. 38, 3, 9, 11, 13).

In sum, it could be speculated that the various plants including the ferns, ginkgos, co-nifers (particularly *Metasequoia*, *Sequoia*, and *Taxodium*, etc.), and the rich flowering plan-tes, were on the "dining tables" of the dinosausrs living in Jiayin and its neighboring eara during the middle-late Late Cretaceous time (Sun, 2019*).

Regarding the gradual dwarfing and expanding distribution of ferns in Mesozoic, Dilcher suggested that it might be related to the large consumption of dinosaurs (Dilcher, 2019**).

4.2　The role of fossil plants in study of the K-Pg boundary in Jiayin

The Cretaceous Paleogene boundary (K-Pg boundary) is a boundary left in the strata about 66 million years ago (66 Ma). During this time, there was large mass extinction of about 70% of the bios, such as terrestrial life represented by dinosaurs and marine life represent-ed by ammonite, which is known as "the fifth mass extinction and recovery of life on earth since Phanerozoic". As far as the study of paleobotany in the Jiayin area of Heilongjiang is concerned, sporopollen fossils retain sufficient evidence, revealing this major geological event that occurred ca. 66 Ma ago in Jiayin and its adjacent areas, especially the great changes of the bios before and after the K-Pg boundary (Sun et al., 2014).

The K-Pg boundary (formerly K/T boundary, KTB) was first proposed by Alvarez et al. in 1980. It was mainly based on the discovery of a large number of iridium (Ir) anoma-lies in the clay of the Cretaceous-Tertiary (K/T) boundary in Europe. After that, interstel-lar materials such as impact quartz were found in the KTB strata near Chicxulub, Yucatan Peninsula, Mexico. Alvarez et al. proposed that the iridium anomaly and shock quartz in the KTB strata were derived from the impact of extraterrestrial bodies on the earth, which caused the extinction of dinosaurs and other creatures. However, many scientists have dif-ferent opinions on this issue (Nichols & Johnson, 2008), they have believed the extinction

---

* Sun G. Mesozoic plants. Text-book for graduates. 2019. Shenyang Normal University.

** Dilcher. Herbivorous dinosaurs and the plants they ate. Lecture in College of Paleontology, Shenyang Normal University in Shenyang, 2019.8.16.

of dinosaurs may be related to the eruption of violent volcanoes (McLean, 1985; Keller et al., 2011). Therefore, the study of the K-Pg boundary has become a "hot spot" in the study of geo-/biological evolution and environmental change at the transition time between the Mesozoic and Cenozoic.

The Cretaceous-Paleogene strata and biota are well developed in Jiayin of Heilongjiang, China, and its neighboring area. A large number of dinosaurs, fish, ostracods, and plant fossils have given testimony to the rich and colorful biological world of Jiayin areas in the period about 86-66 Ma (middle-late Late Cretaceous). The rich plant fossils and other geological remains have also recorded the changes of biological and geological features in this area during 66-60 Ma (Paleocene). The strata along the right bank of Heilongjiang (Amur) River are well exposed and yield very rich fossils, which provides a unique favorable condition for studying the K-Pg boundary here. During 2002-2018, Sun and his international scientific research team carried out cooperative studies on the Late Cretaceous biota and the K-Pg boundary in Jiayin and its neighboring area. The main achievements in the studies include: (1) the age of the Late Cretaceous dinosaur fauna represented by Hadrosaurid is confirmed as the early-middle Maastrichtian, with no finding of dinosaurs in the late Maastrichtian stage (i.e. the Furao Formation time), which is very beneficial for studying the dinosaur extinction in Jiayin and its adjacent areas. That means during the late Maastrichtian (ca. 68-66 Ma) the dinosaurs might be extinct already (Sun et al., 2014); (2) based on the geochemical study of three boreholes (XHY-2005, 2006, 2008) from Xiaoheyan of Jiayin, the research results show that there was no significant "iridium (Ir) anomaly" or any "extraterrestrial material" in the K-Pg boundary strata in Jiayin and its adjacent areas at the time of ca. 66 Ma, from which we can infer that there was no "impact event" during the K-Pg boundary time. The above two important achievements have evidenced that the evolution of the Late Cretaceous biota and their environmental changes in Northeast China and even the whole Northeast Asia have their own unique characteristics which are different from North America and other regions where the "impact events" at the K-Pg boundary time occurred (Sun et al., 2014).

4.2.1 Definition of the K-Pg boundary in Jiayin

During 2002-2011, using comprehensive research methods of paleontology, stratigraphy, isotopic dating, paleomagnetism, and geochemistry, especially with the fine division in high resolution by sporopollen fossils, Sun and his colleagues had defined the K-Pg boundary at the drilling place in Xiaoheyan of Jiayin (129°35′13.5″E, 49°14′53.9″N), and confirmed the strata of 22.00-22.05 m in the borehole XHY-2006 as the stratotype K-Pg boundary (Sun et al., 2011; 2014) (Fig. 39).

The definition of K-Pg boundary is mainly based on the abundant evidence of sporopollen fossils found in the boreholes of XHY-2005, XHY-2006, and XHY-2008. The strata underlying the K-Pg boundary yield the latest Maastrichtian sporopollen of the Late Cretaceous, i.e. the assemblage of *Aquilapollenites conatus-Pseudoaquilapollenites striatus* (Assemblage Ⅴ, from the uppermost Furao Formation). Above the boundary, there is the assemblage of *Triatriopollenites confusus-Aquilapollenites spinulosus* (Assemblage Ⅵ, from the lowest Baishantou Member of Wuyun Formation). The characteristics of sporopollen assemblages Ⅴ and Ⅵ are almost the same as those in the K-Pg boundary strata from Belay Gora in Zeya-Bureya basin, Russia, about 30 km from Xiaoheyan of Jiayin, opposite by the Heilongjiang (Amur) River (Markevich et al., 2011; Sun et al., 2011, 2014). The K-Pg boundary section of Belay Gora is one of the internationally recognized and important candidate standard sections in the world (Nichols & Johnson, 2008). According to the study of Markevich et al. (IBP et al., 2001; Markevich et al., 2005, 2006, 2011), the sporopollen biozones of the K-Pg boundary strata in the Belay Gora were divided into seven assemblages, which is nearly the same as the divisions in the Xiaoheyan of Jiayin. The K-Pg boundary is defined in the interval of the Assemblages Ⅴ and Ⅵ (IBP et al., 2001; Markevich et al., 2006, 2011). The composition of the palynoflora and the correlation of the strata in the Belay Gora (Russia) and Xiaoheyang of Jiayin (China) have documented the accuracy and reliability of the definition of the K-Pg boundary in Jiayin (Sun et al., 2014).

The other evidences associated with the palynological study, include the studies of isotopic dating, paleomagnetism, and geochemistry. The isotopic dating of the volcanic beds (acidic tuff) from the lower Paleocene Baishantou Member of the Wuyun Formation in the Baishantou nearby the Xiaoheyan, overlying the K-Pg boundary, shows the U-Pb zircon age as 64.1 ± 0.7 Ma (Suzuki et al., 2011), which proves that Baishantou Member belongs to the Danian (the early Paleocene). Meanwhile, the correlation dating has been carried out in the coeval volcanic tuff in the Belay Gora section of Russia, and the LA-ICP-MS zircon dating result is 66 ± 1 Ma (Knittel et al., 2013) which has also provided the evidence for the K-Pg boundary correlation between the China and Russia in the same Zeya-Bureya basin. As for the paleomagnetic study, by using magnetic positive and negative pole tests, the K-Pg boundary in the Xiaoheyan of Jiayin should be in the reverse polarity region of c29r (65 ± 0.3 Ma). The K-Pg boundaries determined by paleomagnetic studies of XHY-2006 and XHY-2008 are located near 22.05 m and 23.25 m, which are basically consistent with those determined by the sporopollen fossils (Sun et al., 2011, 2014) (Fig. 40). Moreover, the geochemical analysis shows that there is no iridium anomaly in significant sense in the K-Pg boundary strata in Jiayin.

4.2.2 Importance of fossil plants for the K-Pg boundary research in Jiayin

Although the mega-fossil plants in Jiayin can not play a direct role in the accurate definition of the K-Pg boundary due to the limitation of preservation and boundary determination accuracy, the microfossil plants (spores and pollen) have played a key role in the study of the K-Pg boundary definition in Jiayin.

Since the strata near the K-Pg boundary in Xiaoheyan of Jiayin miraculously preserved the sporopollen fossils from the uppermost Upper Cretaceous to the lowest Paleocene, Markevich et al. identified the existence of K-Pg boundary in the boreholes (XHY-2005, XHY-2006, XHY-2008) in the Xiaoheyan, with the accuracy of "centimeter" and high resolution. The sporopollen assemblage under the K-Pg boundary is the typical *Aquilapollenites stelkii-Pseudointegricorpus clarireticulatus* Assemblage (Assemblage V) in the late Maastrichtian of the Late Cretaceous. Above the K-Pg boundary, the strata yield the assemblage of *Triatriopollenites confusus-Aquilapollenites spinulosus* (Assemblage VI), which is the bottom of Baishantou Member of Wuyun Formation aged in the earliest Paleocene (Danian). The K-Pg boundary is divided between the V/VI sporopollen assemblage zones (Markevich et al., 2011; Sun et al., 2014) (Fig. 41). In the borehole XHY-2006, the K-Pg boundary is located in the interval bed of 22.00-22.05 m, due to its complete sampling rate and the abundance of fossils. The borehole XHY-2006 is selected as the standard section of the K-Pg boundary in Xiaoheyan of Jiayin (as the "Stratotype section"); while the boreholes XHY-2005 (boundary in 19.90-20.30 m) and XHY-2008 (boundary in 23.05-

23.25 m) were selected as the "Paratype sections" (Sun et al., 2014) (Fig. 42).

The detailed study shows that the sporopollen Assemblage V, below the K-Pg boundary, is a typical late Maastrichtian sporopollen assemblage. The spores of this assemblage are mainly *Laevigatosporites*, and the pollen are characterized by the prosperity of Taxodiaceae and Ulmaceae. In the angiosperm pollen, *Aquilapollenites* has 14 species, in which *Aquilapollenites stelkii*, *A. conatus*, *Pseudointegricorpus clarireticulatus*, *Marsypiletes cretacea*, and *Integricorpus bellum* are only limited to the Maastrichtian or the late Maastrichtian in age. The above sporopollen assemblage of Xiaoheyan in Jiayin is consistent with the late Maastrichtian sporopollen assemblage of Zeya-Bureya basin, Sakhalin Islands and Yuri Island, in which the sporopollen assemblages in the latter two areas are supported by marine fossils for the age determinations.

In the sporopollen assemblage Ⅵ above the boundary, some characteristic pollen, such as *Triatriopollenites confusus*, *T. plectosus*, *Aquilapollenites spinulosus* and *A. subtilis* are only found in the early Danian strata. This assemblage is very similar to those of the early Danian assemblages in the Sakhalin islands and Koryak highland, where both have marine bivalves and radiolarians to document the age. It is worth emphasizing that the early Danian spore assemblage yielded in the borehole XHY-2005, shows an obvious "fern spike" characteristics, consistent with those shown in the Zeya-Bureya basin, which provides further evidence for the definition of the K-Pg boundary in Jiayin (Markevich et al., 2011; Sun et al., 2011, 2014).

Besides, the Late Cretaceous-Paleocene mega-plants have played the associated roles for the K-Pg boundary study in Jiayin. For example, the authors' research team found some typical early Paleocene plant fossils, such as *Tiliaephyllun tsagajanicum* (Krysht. et Baik.) Krassilov in the Baishantou Member of Wuyun Formation from the Baishantou site nearby the Xiaoheyan of Jiayin in 2002, which provided the evidence for the determination of the Baishantou Member, above the K-Pg boundary, aged in Danian (Sun et al., 2002; Sun et al., 2005). Moreover, the study of the Santonian-Campanian mega-flora in the Yong'ancun and Tiapinglinchang formations in Jiayin has provided a good foundation for an understanding of the biological and geological changes in the latest Late Cretaceous in the Jiayin area, and provided the circumstantial evidence on the age and stratigraphic position of the Furao Formation which is directly concerned with the K-Pg boundary definition.

Trochodendroides taipinglinchanica

Chapter **5**

Late Cretaceous floral sequences in NE China

As a result of fossil preservation and exposure conditions, there have been not more fossils of the Late Cretaceous plants found in Northeast China so far. The known reliable sites of the Late Cretaceous plants reported from the NE China mainly include the Mudanjiang (Zhang et al., 1980; Zhang, 1981), Jiayin (Zhang, 1984; Tao, 2000; Sun et al., 1995, 2007, 2011, 2014, 2016; Quan, 2006; Quan et al., 2008, 2009; Gong, 2007; Liang & Sun, 2015; Liang et al., 2018a), Songliao Basin (Zheng & Zhang, 1994), and Qitaihe (Sun et al., 2000, 2019), etc. Generally speaking, the study of the Late Cretaceous flora seems to be a weakness in Mesozoic paleobotany in China. Particularly, the successive sequence of the Late Cretaceous flora has not been fully and thoroughly understood.

However, since 2002, the situations on the research of the Late Cretaceous flora has had a turn for the better. The main achievements include the new round of research of the Late Cretaceous flora of Jiayin, in which a lot of discoveries of the Late Cretaceous plants were made, particularly in the study of the Late Cretaceous aquatic angiosperms (Sun et al., 2007, 2016; Quan, 2006; Quan et al., 2008, 2009; Golovneva et al., 2008; Liang et al., 2018a). Moreover, the early Late Cretaceous (Cenomanian) plants (e.g. Platanoid) were newly found in the Qitaihe area, with the associated dating results gained by the authors for the first time (Sun et al., 2019).

The improvement of the research level on the developmental sequence of the Late Cretaceous flora is related to the further study of the Late Cretaceous stratigraphic correlation between the eastern and western Heilongjiang region for looking for fossil energy resources such as oil, in the eastern part of the Heilongjiang Province, as the western part (i.e. the Songliao basin), has already successfully exploited oil and gas resources for a long time. Thus, the study of the Late Cretaceous flora and its successive sequence is potentially significant for the exploration of oil and other fossil fuel resources in the Northeast China, particularly in the eastern Heilongjiang region.

It should be mentioned that in this book we call the staged flora in the Late Cretaceous flora as a "florule", because we regard the Late Cretaceous flora as a whole flora, while the staged florules as the various parts of the whole Late Cretaceous flora. Thus, we have named the known successive sequences composed of the Mudanjiang-, Jiayin- and Yu-Fu florules, respectively.

5.1 Early Late Cretaceous Mudanjiang florule

In this book, the early Late Cretaceous flora of Heilongjiang Province is called "Mudanjiang florule". The main fossil evidence includes the reported Late Cretaceous plant fossils in the Mudanjiang area (Zhang, 1981), the 3 and 4 members of Quantou Formation in the eastern Songliao Basin and Anda area (Zheng et al., 1994), and the Cenomanian plant of Qifenglinchang in eastern Qitaihe (Sun & Zheng, 2000; Sun et al., 2019). The geological age of these fossils is generally the Cenomanian, and some of them may extend to early Turonian. The composition of the florule in this stage is as follows.

Lycopods: *Selaginella suniana* Zheng et Zhang; Filicales: *Cladophlebis* sp., *Onychiopsis psilotoides* (Stocks et Webb) Ward; Conifers: *Thuja heilongjiangensis* Zheng et Zhang; Angiospermae: *Platanus appendiculata*, *P. heilongjiangensis*, Sun et al., *P. pseudiguillelmae*, *P. subnobilis*, *Platanus* sp., *Aralia mudanjiangensis* Zhang, *Dicotylophyllum* sp., etc. In particular, Sun et al. (2019) reported *Platanus heilongjiangensis* in the Qitaihe area, and the isotopic age of the upper volcanic rocks in conformity with the strata bearing the fossil of *P. heilongjiangensis*, is 96.2 ± 1.7 Ma, which appears to give evidence on the fossil and its bearing strata as the Cenomanian (Fig. 43).

The main characteristics of the florule are that fewer ferns and conifers (which may be related to the fossil collection), and more abundant angiosperms, represented by *Platanus*. The early Late Cretaceous florule has only been found in the eastern and central parts of Heilongjiang Province. The florule seems to reflect a hot and dry climate during the stage, and the florule growing environment in the eastern Heilongjiang might be related to the influences from the coastal Pacific Ocean (Zhang, 1981; Sun & Zheng, 2000; Sun et al., 2019).

5.2 Middle of Late Cretaceous florule

The middle of Late Cretaceous florule is referred to the Turonian-Coniacian plant assemblage of the Late Cretaceous flora in the eastern Heilongjiang region. The outlook of the Santonian assemblage represented by the plants from the Yong'ancun Formation in Jiayin is clear. While the Turonian is the latter of the early Late Cretaceous on the stratigraphic scheme, the "middle florule" of the Late Cretaceous studied in this book actually refers to the floral assemblage later than the early Late Cretaceous Mudanjiang florule, and earlier than the middle-late Late Cretaceous Jiayin florule, i.e. the florule aged in about Turonian-Coniacian.

Up to now, the Turonian-Coniacian florule in the Heilongjiang region has still been unknown in its floral outlook. The strata yielding the florule are distributed in the Qingshankou Formation (ca. 90.0-86.3 Ma) in the eastern margin of the Songliao basin, which is the western part of the Heilongjiang region. Although there is no reliable record of the mega-fossil plants, some sporopollen fossils have been found (Xi et al., 2019). Thus searching and studying the mega-plant fossils of the Turonian-Coniacian stages, appear to be the main task for the further study of the Late Cretaceous successive sequence in the Heilongjiang region.

As for the aquatic angiosperm fossils of *Trapa angulata* (Newb.) Brown from the Yaojia Formation in the eastern margin of the Songliao Basin (Zheng et al., 1994), the fossils have been identified as the aquatic angiosperms *Quereuxia angulata* (Newb.) Krysht. They may belong to the middle-late Late Cretaceous florule (i.e. the Jiayin florule). The age of the Yaojia Formation (86.3-84.2 Ma) may be roughly equivalent to the Yong'ancun Formation in age (Sun et al., 2014; Xi et al., 2019).

5.3 Middle-late Late Cretaceous Jiayin florule

The middle-late Late Cretaceous Jiayin florule in the Heilongjiang region is mainly represented by the Yong'ancun floral assemblage of *Parataxodium - Nelumbo* (Santonian) and the Taipinglinchang assemblage of *Metasequoia - Trochodendroides - Quereuxia* (Campanian). More than 27 species belonging to 24 genera have been found in Yong'ancun assemblage, including *Equisetum* sp., *Asplenium dicksonianum* Heer, *Arctopteris* sp., *Cladophlebis* sp., *Gleichenites* sp.; *Ginkgo adiantoides* (Ung.) Heer, *G. pilifera* Samylina; *Cupressinocladus sveshnikovae* Ablajev, *Metasequoia disticha* (Heer) Miki, *Sequoia* sp., *Parataxodium* sp., *Elatocladus* sp. 2; *Dalembia jiayinensis* Sun et Golovneva, *Menispermites* sp., *Trochodendroides arctica* (Heer) Berry, *Nyssidium arcticum* (Heer) Iijinskaja, *Platanus* sp., *Viburnophyllum* sp., *Dicotylophyllum* sp., *Quereuxia angulata* (Newb.) Krysht., *Cobbania corrugata* (Lesq.) Stockey et al., *Nelumbo jiayinensis* Liang et al., etc.

The Taipinglinchang assemblage is composed of more than 38 species of 30 genera, including *Thallites* sp.; *Equisetum* sp., *Asplenium dicksonianum* Heer, *Cladophlebis* sp.; *Ginkgo adiantoides* (Ung.) Heer, *G. pilifera* Samylina; *Taxodium olrikii* (Heer) Brown, *Metasequoia disticha* (Heer) Miki, *Sequoia* sp., *Pityophyllum* sp., *Pityospermum* sp., *Glyptostrobus* sp., *Larix* sp., *Elatocladus* spp. 1, 2; *Araliaephyllum?* sp., *Arthollia orientalis* (Zhang) Golovneva, *A. tschernyschewii* (Kostanov) Golovneva, Sun et Bugdaeva, *Celastrinites kundurensis* Gol., Sun et Bugd., *Platanus densinervis* Zhang, *P. sinensis* Zhang, *Platanus* sp., *Trochodendroides arctica* (Heer) Berry, *T. taipinglinchanica* Gol., Sun et Bugd., *T. lanceolata* Gol., *T. microdentatus* (Newb.) Krysht., *Viburnum* cf. *contortum* Berry, *Viburnophyllum* sp., *Quereuxia angulata* (Newb.) Krysht., *Cobbania corrugata* (Lesq.) Stockey et al., etc.

The above two assemblages are characterized by the following: ① the proportion of angiosperms in the composition increases significantly, and is dominant in the florule (up to 50%); although some earlier elements in the Late Cretaceous (e.g. *Platanus*, *Araliaephyllum*, and *Menispermites*) still exist, the florule has some important elements of the middle-late Late Cretaceous angiosperms, such as *Dalembia*, *Artholia*, and *Celastrinites* (Fig. 44);

② among gymnosperms, *Metasequoia* and *Sequoia* appeared for the first time, indicating that the floral composition is gradually transiting to the Cenozoic; ③ warm and humid plants still account for a large proportion (e.g. ginkgos and ferns); ④ a large number of aquatic angiosperms, such as *Quereuxia*, *Cobbania*, and *Nelumbo*, occurred for the first time, reflecting the warm temperate climate environment with abundant water supply.

5.4 The latest Late Cretaceous Yu-Fu florule

The latest Late Cretaceous (Maastrichtian) florule is represented by the plants (mainly shown by sporopollen) from the Yuliangzi Formation (early-middle Maastrichtian) and the Furao Formation (late Maastrichtian) in the Jiayin area. The authors newly name it as the "Yu-Fu florule", in which "Yu" means Yuliangzi Formation, while the "Fu" represents the Furao Formation. The age of the florule is documented by the sporopollen fossils as the whole Maastrichtian in total. Although no mega-fossil plants have been reported so far, the abundant and unique sporopollen fossils provide reliable evidence (see 2.1.2).

5.5 Outline of Late Cretaceous floral sequences in NE China

Although the floral developmental sequence of the Late Cretaceous flora in Northeast China is still not perfect in demonstration, according to the known achievements of the studies of the Late Cretaceous flora mentioned above, the outline of the sequences for the floral development in the NE China can be summarized as follows.

① The early developmental stage is represented by the early Late Cretaceous Mudanjiang florule which is mainly found from the Mudanjiang-Qitaihe area of the Heilongjiang region, and dominated in the Cenomanian age.

② The middle (or early-middle) developmental stage is represented by the "middle" florule of the Late Cretaceous, which is mainly aged in the Turonian-Coniacian. The composition of this florule is still under study.

③ The middle-late developmental stage is represented by the middle-late Late Cretaceous Jiayin florule which is aged in the Santonian-Campanian. This stage has been already studied in detail (see 2.1.1).

④ The latest developmental stage is represented by the latest Late Cretaceous Yu-Fu florule which is aged in the Maastrichtian. The floral characters of this stage are demonstrated mainly by the sporopollen, while the mega-fossil plants of the florule await further study.

According to the authors' study, the main recent advances in the study of the Late Cretaceous flora in Jiayin of Heilongjiang can be summarized as follows:

1. The features of the early Late Cretaceous (Cenomanian) and the middle-late Late Cretaceous (Santonian-Maastrichtian) floristic assemblages in the eastern Heilongjiang region have been preliminarily and successfully studied and some well recognized, although the Turonian-Coniacian florule remains to be studied.

2. The characteristics of the floristic evolution or development of the Late Cretaceous flora in the Heilongjiang region are highlighted by the angiosperms gradually flourishing, especially for the aquatic flowering plants, in which, the angiosperms have occupied a dominant position in this whole flora (up to 50%, even more) since the Santonian or Santonian-Campanian.

Table 7   Outline of developmental sequences of Late Cretaceous floras in NE China

Horizons			Stage	Ma	florule name	Main characteristics	Note
K₂	U	6	Maastrichtian	66.0 72.1	Yu-Fu Florule	Typical sporopollen of late Maastrichtian	Abundant dinosaurs in early-middle substages
	M	5	Campanian	84.2	Jiayin Florule	*Dalembia*, *Artholia*, *Trochodendroides*, *Metasequoia*, with abundance of aquatic flowering plants. Angiosperms up to 50%	
		4	Santonian	86.3			Herbirovous dinosaurs
	L	3	Coniacian	89.8	?		
		2	Turonian	93.9			
		1	Cenomanian	100.5	Mudanjiang Florule	Abundant Platanoid	96.2 ± 1.7 Ma for reference

3. Based on the evidence of the flora, the paleoclimatic characteristics and changes seem to be as follows: in the early stage （Cenomanian） of the Late Cretaceous in Heilongjiang, it was relatively hot; while during the middle-late Late Cretaceous it became warm or warm-temperate and more humid, with some seasonal changes.

As for the main problems and future work on the study of Late Cretaceous flora, particularly in its floral developmental sequence in the Heilongjiang region, the authors would like to suggest as follows.

1. The study on the floral assemblage of the Turonian-Coniacian stages is urgently needed, and it is therefore necessary that the study of floral evolutionary or developmental sequences of the Late Cretaceous flora of the NE China needs to be further carried out.

2. For more evidence on the ages of the plant-bearing strata in the Upper Cretaceous in Heilongjiang region, particularly for the Turonian-Coniacian strata, it is important to find more volcanic rocks and make their isotopic dating.

Nyssidium arcticuma

Chapter **6**

Science popularization with paleobotanical study in Jiayin

In the past 20 years, with the support from state and Heilongjiang provincial governments, and along with the achievements made in geology and paleontology, highlighted by dinosaur and plant fossils and the K-Pg boundary, the popularization and application of Geology and Paleontology in Jiayin have been revived. Many scientists and students from Russia, Germany, USA, UK, Belgium, France, Japan, and many other countries came to Jiayin for seeing the fossil miracle of Jiayin.

The recent achievements in the popularization of Geology and Paleontology in Jiayin mainly include: (1) the establishment of the first Paleontologist Sculpture Garden of Jiayin in China; (2) the establishment of the first national dinosaur Geopark and dinosaur museum in the border area between China and Russia; and (3) the establishment of the first non-marine K-Pg boundary point in China. Besides using the dinosaurs like the "leading star", publicity work and the scientific popularization of Palaeobotany and Palynology in Jiayin have been unique and received extensive praise.

6.1 Sculpture Garden of Paleontologists in Jiayin

To commemorate and thank Chinese and foreign geoscientists for their contributions to Jia-yin, Heilongjiang, the Jiayin County Government established the first "Sculpture Garden of Paleontologists" in the National Dinosaur Geopark of Jiayin in 2006. Around 30 m long dinosaur skeleton model in the Garden, stand 24 bronze statues of internationally famous scientists who have made contributions to the geoscientific studies in Jiayin of Heilongji-ang, including seven famous Chinese academicians: Li S. G. (J. S. Lee), Yang Z. J. (C. C. Young), Si X. J. (H. C. Sze), Yang Z. Y., Li X. X., Gu Z. W., and Hao Y. C., and Profes-sors Dong Z. M., Zhao X. J., and Sun G. The famous foreign scientists include eight acade-micians, T. H. Huxley, R. Owen, de P. T. Chardin, A. N. Riabinin, A. N. Kryshtofovich, J. Ostrom, B. Chalener, D. L. Dilcher, M. Akhmetiev, and Professors V. Markovich, A. R. Ashraf, K. Johnson, and P. Curry (Fig. 45).

Besides the famous dinosaur experts at home and abroad, among the sculptures, there are many palaeobotanists such as Si X. J., Li X. X., Sun G. (China), Kryshtofovich A. N., M. Akhmetiev. (Russia), D. L. Dilcher, K. Johnson (USA), and palynologists V. Markovich (Russia), and A. R. Ashraf (Germany).

These vivid bronze statues are engraved with love and respect of Jiayin people for Sci-ence and scientists, and have left a deep impression and influence on the people, especially the youth. Since 2006, many tourists at home and abroad have traveled thousands of miles to visit this statue garden, where they learn and feel the stories and spirit of dedicated scien-tists (Fig. 46).

6.2 Fossil plants in Jiayin Dinosaur Museum

The second achievement in Jiayir's science popularization is the establishment of China's first paleontological museum in the Sino-Russian border area, i.e. Jiayin Dinosaur Museum, associated with the National Dinosaur Geopark of Jiayin. In this Museum, apart from *Mand-schurosaurus* and other hadrosaurid fossils as the "stars" in the main exhibits, there is also a high-level exhibition displaying the plant fossils from the Upper Cretaceous and Paleocene in Jiayin.

The display of the Late Cretaceous plant fossils includes the well-known fossils of *Nulumbo jiayinensis*, *Dalembia*, *Quereuxia*, and *Cabbonia*; the co-evolution of dinosaurs with plants in the Late Cretaceous and their paleoecological reconstructions are also vivid and attractive (Fig. 47).

6.3 Celebration activities related to Geology and Paleontology

Between 2002 and 2019 the paleontological research had been ceaselessly pushed to a climax, in which the paleobotanical and palynological researches had played a huge driving force. It is worth mentioning that the study of the mega-plants and microplant fossils have played very important roles in the research of the Late Cretaceous biota, and the divisions of the Upper Cretaceous and Paleocene strata in Jiayin, particularly the palynological study, has contributed a lot to the definition of K-Pg boundary in Jiayin and helped in defining the dinosaur age in Jiayin, which in turn helped us to think about the dinosaur extinction in China and even in Northeast Asia. In nearly 20 years in the past, about ten large-scale celebrations or science popularization activities with the topics on fossils had been held in Jiayin (Figs. 48, 49).

In September 2002, the "First International Symposium on Cretaceous Biota and K-T Boundary" was held in Jiayin. Scientists from China, Russia, Germany, USA, UK, Japan, and ROK arrived at Jiayin, renewing the understanding on the part of the county government leaders and the broad masses. A series of new development visions, such as "creating dinosaur brand and driving Jiayin's economic development", was gradually breeding, blooming, and fruiting in Jiayin. Then, with the strong support of the state and the provincial governments of Heilongjiang, the scientific activity to protect and popularize the knowledge of dinosaurs and other fossils continued to drive the economic development of Jiayin. The people, especially teenagers, have not only improved their scientific knowledge from fossils but also seen the hope of Jiayin in promoting cultural and economic development by using fossils. In 2011, Jiayin's popularization and celebration of fossil paleontology set off a new upsurge, and during August 21-24, the field investigation and celebration of the K-Pg boundary point were conducted at the Xiaoheyan village of Jiayin. A festive air prevailed in Jiayin when more than 150 scientists from 16 countries around the world attended the ceremony. Experts at home and abroad expressed sincere congratulations to the Jiayin's "K-Pg boundary Point" being chosen as one of the international candidate sites (point No. 95) (Fig. 48, 7-9).

The new climax of Jiayin fossil publicity and science popularization activities is in 2017-2019. In Sept. 2-3, 2017, the celebration on "The 115th Anniversary of Discovery of Dinosaurs in China and the First Jiayin Fossil Protection Forum" was held in Jiayin. More than 100 experts from 30 provinces and cities in China, as well as teachers and students from University of Bonn, Germany, gathered in Jiayin. Through this meeting, the participants not only exchanged the experience of geological and paleontological popularization work, but also made the Fossil Protection Form of Jiayin as one of the "national brands" (Fig. 49, 1-3). In Aug.18-21, 2019, Jiayin held the "Int'l symposium on Cretaceous biota and K-Pg boundary in Jiayin, Heilongjiang, and the second Jiayin Fossil Protection Forum", which is the largest international conference in Heilongjiang Province for the recent years. More than 180 participants from 16 countries, including China, USA, Germany, UK, Russia, Belgium, Romania, Japan, DPRK, ROK, Thailand, India, Kyrgyzstan, Brazil, etc. attended this conference. Scientists at home and abroad highly appraised the new progress in the study of K-Pg boundary in Jiayin, China, and strongly recommended the K-PgB point in Jiayin as China's national standard (Fig. 49, 4-10). These activities not only promoted the publicity work and scientific popularization of geological and paleontological achievements in Jiayin but also greatly strengthened the fossil protection and boosted the development of geoscientifical cultur tourism in Jiayin.

Acknowledgements

The author would like to thank all the experts of the int'l research team in Jiayin (2002-2011), particularly the paleobotanists M. Akhmetiev, D. L. Dilcher, K. Johnson, H. Nishida, L. Golovneva, T. Kodrul, and Bugdaeva E., as well as Sun C. L., Sun Y. W., and C. Quan; and palynologists V. Markevich, A. R. Ashraf, D. Nichols, I. Harding, and T. Kezina for their great dedication and help (Fig. 50).

Thanks are also extended to the experts of the int'l research team Dong Z. M., P. Godfroit, Yu & I. Bolotsky, Suzuki S., Terada K., Tsukagoshi M., and M. Tekleva, as well as Yang H. X., Ge W. C., Chen Y. J., Gong F. H., and Feng Y. H. Particularly to Academicians Li T. D. and Liu J. Q., and Prof. Wang Y. D. for their kind support for publishing this book, and to Prof. Wan X. Q. and Dr. Xi D. P. for kindly providing references; to Dr. Wang S. P. (Editor in Chief) and Dr. Wu H. L. (SSTE) for their kind support and editing of this book.

The research work concerning this book has been supported by the projects NSFC - 3022130698, 40842002 and 41602015; Projects 2015FY310100 (MOST), DD 20160120-04 (CGS); Project "111"(MOEC, in JU.), Key-Lab of Evolution of Past Life in NE Asia (MNRC, in SYNU), Key-Lab of Evolution of Past Life Evolution & Environment in NE Asia (MOEC, in JU), and Proj.183117 (State Key-Lab of Modern Paleontology & Stratigraphy, CAS).

Finally, the author wishes to express sincere thanks to the leaders and colleagues of the Dept. Natural Resources of Heilongjiang, Geological Museum of Heilongjiang, the 6th Geological and Exporting Institution, Geological Survey No. 1 of Heilongjiang, Yichun Municipal Government and Jiayin County Government, and the colleagues of RCPS (JU), PMOL and CP (SYNU) for their kind support and help.

References

Ablaev A G. 1974. *Late Cretaceous Flora of Sikhote-Alin and Its Significance for Stratigraphy*. Novosibirsk: Nauka Siberian Branch, 1-179. (in Russian)

Abramova A L. 1983. Conspectus of the moss flora of the People's Republic of Mongolia. Biological resources and natural conditions of the People's Republic of Mongolia, 17. Leningrad: Nauka, 1-221. (in Russian)

Akhmetiev M A. 2004. Biosphere crisis at the Cretaceous-Paleogene boundary. In: Sun G, Sun Y W, Akhmetiev M A, et al (eds). *Proceeding of the 3rd symposium on Cretaceous Biota and K/T boundary in Heilongjiang River area*, Changchun, 7-16.

Alvarez L W, Alvarez W, Asaro F, et al. 1980. Extraterrestrial cause for the Cretaceous-Tertiary extinction. *Science*, 208: 1095-1108.

Bell W A. 1949. Uppermost Cretaceous and Palaeocene floras of western Alberta. *Can. Geol. Surv. Bull.*, 13: 1-231.

BGMRHP (Bureau of Geology and Mineral Resources of Heilongjiang Province). 1993. *Regional Geology of Heilongjiang Province*. Beijing: Geological Publishing House, 1-734. (in Chinese)

Bolotsky I. 2013. Tyrannosaurid dinosaurs (Coelusauria) from Upper Cretaceous Amur/Heilongjiang River area. Master Thesis of Jilin University, China, 1-88.

Bolotsky Y, Godefroit P. 2004. A new hadrosaurine dinosaur from the Late Cretaceous of Far Eastern Russia. *Journal of Vertebrate Paleontology*, 24: 354-368.

Chen P J. 2000. Comments on the classification and correlation of non-marine Jurassic and Cretaceous of China. *J. Stratigraphy*. 24(2): 114-119. (in Chinese with English abstract)

Dilcher D L. 1974. Approaches to the identifications of angiosperm leaf remains. *Bot. Rev. (Lancaster)*, 40 (1): 1-157.

Dilcher D L. 2000. Toward a new synthesis: Major evolutionary trends in the angiosperm fossil record. *PNAS*, 97: 7030-7036.

Dong Z M, Zhou Z L, Wu S Y. 2003. Note on a Hadrosaur footprint from Heilongjiang River area of China. *Vertebrata Palasiatica*, 41(4): 324-326. (in Chinese with English abstract)

Dong Z M. 2009. Dinosaur in Asia. Kunming: Yunnan Science & Technology Press, 1-287. (in Chinese with English abstract)

Doyle J A, Hickey L J. 1976. Pollen and leaves from the mid-Cretaceous Potomac Group and their bearing on early angiosperm evolution. In: Beck C B (ed). Origin and Early Evolution of Angiosperms. New York: Columbia University Press, 139-206.

Duan J Y, An S L. 2001. Early Cambrian Siberian Eauna from Yichun of Heilongjiang Province. *Act. Pal-*

aeont. Sin., 40(3): 362-370. (in Chinese with English abstract)

Feng G P, Li C S, Zhilin S, et al. 2000. *Nyssidium jiayinense* sp. nov. (Cercidiphyllaceae) of the Early Tertiary from north-east China. *Bot. J. Linn. Soc.*, 134(3), 471-484.

Fontain W M. 1889. The Potomac or Younger Mesozoic flora. *US Geological Survey Monograph*, 15:1-377.

Godefroit P, Hai S, Yu T, et al. 2008. New hadrosaurid dinosaurs from the uppermost Cretaceous of northeastern China. *Acta Palaeontologica Polonica*, 53: 47-74.

Godefroit P, Lauters P, Itterbeeck J V, et al. 2011. Recent advances on the study of hadrosaurid dinosaurs in Heilongjiang (Amur) River area between China and Russia. *Global Geology*, 13(4): 160-191.

Golovneva L B. 1994. Maastrichtian-Danian floras of Koryak Upland. *Proc. Komarov Bot. Inst. RAS.*, 13: 1-146. (in Russian with English summary)

Golovneva L B. 2005. Phytostratigraphy and Evolution of Albian-Campanian Flora in Siberia. In: *Proceedings of* II *All-Russia Conference on the Cretaceous System of Russia: Problems of Stratigraphy and Paleogeography, St. Petersburg*. St. Petersburg: S. -Peterb. Gos. Univ., 177-197.

Golovneva L B, Sun G, Bugdaeva E. 2008. Campanian flora of the Bureya River Basin(Late Cretaceous of the Amur Region). *Paleontological Journal*, 42(5): 554-567.

Gong F H. 2007. Late Cretaceous Ginkgo from Jiayin of Heilongjiang, China. Master Thesis of Jilin University. 1-80. (in Chinese with English abstract)

Guo S X. 1984. Cretaceous plants from Songliao basin. *Act. Palaeont. Sin.*, 23(1): 85-90. (in Chinese with English abstract)

Guo S X. 1986. Cretaceous Flora features and evolution from China and the Northern hemisphere. *Act. Palaeont. Sin.*, 25(1): 31-45. (in Chinese with English abstract)

Guo Z, Jia H M, Guan H T, et al. 2014. Determination of a new Late Cretaceous volcano group in Jianan of Liaoyuan. *Global Geology*, 33(4): 787-792.

Hao Y C, Su D Y, Yu J X, et al. 2000. S*tratigraphy of China. Cretaceous System*. Beijing: Geological Publishing House. 1-124. (in Chinese)

Haq B U, Hardenbol J, Vail P R. 1987. Chronology of fluctuating sea level since the Triassic. *Science*, 235: 437-455.

He S A, Yin G, Pang Z J. 1997. Resources and prospects of *Ginkgo biloba* in China. In: Hori T, et al. (eds). Ginkgo boiloba, *A Global Treasure: From Biology to Medicine*. Tokyo: Springer, 373-384.

Heer O. 1878. Beitraege zur Fossilen Sibiriens und des Amurlandes. *Mem. Acad. Imp. Sci. Saint-Petersb. Ser.*, 7(25): 1-61.

Herman A B, Lebedev E L. 1991. Cretaceous stratigraphy and flora of the northwestern Kamchatka Peninsula. *Trans. Geol. Inst. Acad. Sci. USSR*, 468. (in Russian)

Herman A B. 2002. Late Early-Late Cretaceous floras of the North Pacific Region: Florogenesis and early angiosperm invasion. *Rev. Paleobot. Palyn.*, 122: 1-11.

Herman A B. 2011. Albian-Paleocene flora of the North Pacific region. *Trans. Geol. Inst*. 592. Moscow: GEOC, 1-280.

Hu X M. 2004. Greenhouse climate and ocean during the Cretaceous. *Geology in China*, 31: 442-448. (in Chinese)

IBP, et al. 2001. *Flora and Dinosaurs at the Cretaceous-Paleogene Boundary of Zeya-Bureya Basin*. Vladivostok: Dalnauka, 1-159.

Johnson K R. 2002. Megaflora of Hell Creek and lower Fort Union Formations in the Western Dakotas: Vegetational response to climate change, the Cretaceous-Tertiary boundary event, and rapid marine transgression. In: Hartman J H, et al (eds). *The Hell Creek Formation and the Cretaceous-Tertiary Boundary in the Northern Great Plains: An Integrated Continental Record of the End of the Cretaceous*. Boulder: Geological Society of America, 361: 329-390.

Keller G B, Bhowmick P K, Upadhyay H, et al. 2011. Deccan volcanism linked to the Cretaceous-Tertiary boundary mass extinction: New evidence from ONGC Wells in the Krishna-Godavari Basin. *J. Geol. Soc. India*, 78: 399-428.

Knittel U, Suzuki S, Akhmetiev M A, et al. 2013. 66 ± 1 Ma single zircon U-Pb date confirms the location of the non-marine K-Pg boundary in the Amur/Heilongjiang River area (Russia, China). *Neues Jahr. Geol. Palaont.*, 270(1): 1-11.

Krassilov V A. 1976. *Tsagajan flora of Amur Region*. Moscow: Nauka, 1-191. (in Russian)

Krassilov V A. 1979. *Cretaceous Flora of Sakhalin*. Moscow: Nauka, 1-185. (in Russian)

Kryshtofovich A N. 1953. Some enigmatic plants of the Cretaceous flora and their phylogenetic significance. *Paleontol. Strat. Trudy VSEGEI.*, 18–30. (in Russian)

Kryshtofovich A N, Baikovskaya T N. 1966. Upper Cretaceous Tsagayan flora in the Amur Region. In: Kryshtofovich A N. Selected Papers. M.: Akad. Nauk SSSR, 3: 184-320. (in Russian)

Lebedev E L. 1974. Albian Flora and Lower Cretaceous Stratigraphy of West Priokhotie. Moscow: Nauka, 1-147. (in Russian)

Lebedev E L, Herman A B. 1989. A new genus of Cretaceous angiosperm—*Dalembia*. *Review of Paleobotany and Palynology*, 59: 77-91.

Lesquereux L. 1878. Contributions to the fossil flora of the Western Territories. Part 2. The Tertiary Flora. Report of the United States Geological Survey of the Territories, 7: 366.

Li G, Chen P J, Wan X Q, et al. 2004. Stratotype of the basal boundary of the Nenjiang Stage, Cretaceous. *Journal of Stratigraphy*, 28(4): 297-299, 335. (in Chinese with English abstract)

Li J T, Zhang J, Yang T N, et al. 2009. Crustal tectonic division and evolution of the southern part of the North Asian Orogenic Region and its adjacent areas. *Journal of Jilin University (Earth Science Edition)*. 39(4): 584-605.

Li X X. 1959. *Trapa? microphylla* Lesq., the first occurrence from the Upper Cretaceous Formation in China. *Act. Palaeont. Sin.*, 7(1): 33-40. (in Chinese with English abstract)

Liang F, Sun G. 2015. New discovery of aquatic angiosperm *Cobbania* from the Upper Cretaceous Yong'ancun Formaion in Jiayin of Heilongjaing. *Global Geology*, 34(1): 1-6. (in Chinese with English abstract)

Liang F, Sun G, Yang T, et al. 2018a. *Nelumbo jiayinensis* sp. nov. from the Upper Cretaceous Yong'ancun Formation in Jiayin of Heilongjiang, Northeast China. *Cretaceous Research*, 84: 134-140.

Liang F, Wu Q, Yuan L F, et al. 2018b. Cuticles of aquatic angiosperm *Quereuxia* from Upper Cretaceous of Jiayin of Heilongjiang. Abstracts of the 29th Annual Scientific Conference of Paleontological Society of China. Zhengzhou of Henan, Sept. 2018. 171. (in Chinese)

Liu M L. 1990. Upper Cretaceous and Tertiary palyno-assemblage sequences in Northeast China. *Journal of Stratigraphy*, 14(4): 277-285. (in Chinese with English abstract)

Markevich V S, Golovneva L B, Bugdaeva E V. 2005. Floristic Characterization of the Santonian-

Campanian Deposits of the Zeya-Bureya Basin (Amur Region). In: Proceedings of International Conference on the Current Problems in Paleofloristics, Paleophytogeography, and Phytostratigraphy, Moscow, May 17-18, 2005. (in Russian)

Markevich V S, Sun G, Ashraf A R, et al. 2006. The Maastrichtian-Danian palynological assemblages from Wuyun of Jiayin nearby the Heilongjiang(Amur) River. In: Yang Q, et al (eds). Ancient life and modern appraches. Abstracts of the 2nd IPC, Beijing, 526-527.

Markevich V S, Ashraf A R, Nichols D, et al. 2008. The most important taxa for correlation of the Santonian to Danian deposits of Far East. The 2nd Workshop on the K-T boundary in Jiayin of Heilongjiang, China, Changchun, Nov. 16, 2008. 23.

Markevich V S, Bugdaeva E V, Sun G. 2009. Palynoflora of Wulaga dinosaur site in Jiayin (Zeya-Bureya Basin, China). *Global Geology*, 13(3): 117-121.

Markevich V S, Bugdaeva E V, Ashraf A R, et al. 2011. Boundary of Cretaceous and Paleogene continental deposits in Zeya-Bureya Basin, Amur(Heilongjiang)River region. *Global Geology*, 14(3): 144-159.

McLean D M. 1985. Deccan traps mantle degassing in the terminal Cretaceous marine extinctions. *Cretaceous Research*, 6: 235-259.

Miki S. 1941. On the change of flora in Eastern Asia since Tertiary Period (I). The clay or lignite beds flora in Japan with special reference to the *Pinus trifolia* beds in Central Hondo. *Jap. J. Bot.*, 11(3): 237-304.

Mosbrugger V, Utescher T. 1997. The coexistence approach: a method for quantitative reconstructions of Tertiary terrestrial palaeoclimate data using plant fossils. *Palaeogeography, Palaeoclimatology, Palaeoecology*, 134: 61-86.

Mosbrugger V. 1999. The nearest living relative method. In: Jones Y P, Rowe N P (eds). *Fossil Plants and Spores: Modern Techniques*. London: Geological Socienty, 261-265.

Newberry J S. 1861. Geological Report, fossil plants. In: Ives (ed). Report upon the Colorado River of the west explored in 1857 and 1858 by Lieutenant Joseph C. Ives. Corps of Topographical Engineers, Office of Explorations and Surveys. GPO, Washington, D.C., 129-132.

Newberry J S. 1898. Later extinct floras of North America. *US Geological Survey Monograph*, 35: 1-151.

Nichols D, Johnson K. 2008. *Plants and the K-T Boundary*. Cambridge: Cambridge University Press, 1-292.

Quan C. 2006. Late Cretaceous flora and strata from Jiayin along with Heilongjiang River area. Ph.D. Thesis. Jilin University, 1-206. (in Chinese)

Quan C, Sun G. 2008. Late Cretaceous aquatic angiosperms from Jiayin, Heilongjiang, Northeast China. *Acta Geologica Sinica*, 82(6): 1133-1140.

Quan C, Sun C L, Sun Y W, et al. 2009. High resolution estimates of paleo-CO_2 levels through the Campanian (Late Cretaceous) based on *Ginkgo* cuticles. Cretaceous Research, 30: 424-428.

Retallack G J. 2001. A 300-million-year record of atmospheric carbon dioxide from fossil plant cuticles. *Nature*, 411: 287-290.

Riabinin A N. 1930. On the age and fauna of the dinosaur beds on the Amur River. *Mem. Russian Paleont. Soc.*, 59(2): 41-51.

Russell D. 1970. Tyrannosaurs from the Late Cretaceous of Western Canada. *Palaeontology*, 12(1): 1-34.

Samylina V A. 1963. The Mesozoic flora of the lower course of the Aldan River. *Paleobotanica*, Ⅳ. Moscow: Nauka, 59-139. (in Russian with English summary)

Samylina V A. 1967. The Mesozoic flora of the area to the west of the Kolyma River (the Zyrianka coal-basin) . Ginkgoales, Coniferales. General Chapters. *Paleobotanica* Ⅵ, 133-175. (in Russian with English abstract)

Samylina V A. 1988. *Arkagalinskaya Stratoflora of Northeast Asia*. Leningrad: Nauka, 1-131. (in Russian with English abstract)

Spicer R A, Herman A B. 2001. The Albian-Cenomanian flora of the Kukpowruk River, western North Slope, Alaska: Stratigraphy, palaeofloristics, and plant communities. *Cretaceous Research*, 22 (1): 1-40.

Stockey R A, Rothwell G W, Johnson K. 2007. *Cobbania corrugata* gen. et comb. nov. (Araceae): A floating aquatic monocot from the Upper Cretaceous of western North America. *Amer. J. Bot.*, 94: 609-624.

Sun G, Cao Z Y, Li H M, et al. 1995. Cretaceous floras. In: Li X X(eds). *Fossil Floras of China through the Geological Ages*. Guangzhou: Guangdong Science & Technology Press, 310-344.

Sun G, Zheng S L. 2000. New proposal on division and correlation of Mesozoic stratigraphy for NE China. *Journal of Stratigraphy*, 24(1): 60-64. (in Chinese with English abstract)

Sun G, Zheng S L, Wang X F, et al. 2000. Subdivision of developmental stages of early angiosperms from NE China. *Act. Palaeont. Sin.*, 39 (Sup.): 186-199.

Sun G, Akhmetiev M A, Dong Z M, et al. 2002. In search of the Cretaceous-Tertiary boundary in the Heilongjiang River Area of China. *J. Geosci. Res. NE Asia*, 5(2): 105-113.

Sun G, Sun C L, Dong Z M, et al. 2003. Preliminary study of the Cretaceous- Tertiary boundary in Jiayin of the Heilongjiang River area of China. *Global Geology*, 22(3): 8-14. (in Chinese with English abstract)

Sun G, Quan C, Sun C L, et al. 2005. Some new knowledge on subdivisions and age of Wuyun Formation in Jiayin of Heilongjiang, China. *Journal of Jilin University* (Earth Science Edition), 35(2): 137-142. (in Chinese with English abstract)

Sun G, Akhmetiev M A, Golovneva L B, et al. 2007. Late Cretaceous plants from Jiayin along Heilongjiang River, Northeast China. *Courier Forschungsinstitut Senckenberg*, 258: 75-83.

Sun G, Akhmetiev M, Markevich V, et al. 2011. Late Cretaceous biota and the Cretaceous-Paleocene (K-Pg) boundary in Jiayin of Heilongjiang, China. *Global Geology*, 14(3): 115-143.

Sun G, Dong Z M, Akhmetiev M, et al. 2014. *Late Cretaceous-Paleocene Biota and the K-Pg Boundary from Jiayin of Heilongjiang*, China. Shanghai: Shanghai Science, Technology and Education Publishing House. 1-194.

Sun G, Golovneva L, Alekseev P, et al. 2016. New species *Dalembia jiayinensis* (Magnoliopsida) from the Upper Cretaceous Yong'ancun Formation, Heilongjiang, northern China. *Cretaceous Research*, 67: 8-15.

Sun G, Wang L X, Sun Y W, et al. 2018. *Basic Course on Fossil Protection*. Shanghai: Shanghai Science, Technology and Education Publishing House, 1-159.

Sun G, Kovaleva T, Liang F, et al. 2019. A new species of *Platanus* from the Cenomanian (Upper Cretaceous) in eastern Heilongjiang, China. *Geosciences Frontiers*, 7: 8-15.

Suzuki S, Sun G, Ulrich K, et al. 2011. Radiometric Zircon Ages of a Ruff Sample from the Baishantou

Member of Wuyun Formation, Jiayin: A Contribution to the Search for the K-T boundary in Heilongji-ang River Area, China. *Acta Geologica Sinica*, 85(6): 1351-1358.

Sveshnikova I N. 1963. Atlas and key for the identification of the living and fossil Sciadopity-aceae and Taxodiaceae based on the structure of the leaf epiderm. *Paleobotanica* Ⅳ. Moscow: Nauka, 207-229. (in Russian with English abstract)

Sze X J, Li X X, et al. 1963. *Fossil Plants of China* (2). *Mesozoic Plants*. Beijing: Sciences Press, 1-429. (in Chinese with English abstract)

Tao J R. 2000. *The Evolution of the Late Creatceous-Cenozoic Floras in China*. Beijing: Sciences Press, 1-282. (in Chinese)

Taylor T N, Taylor E L, Krings M. 2009. Paleobotany, the biology and evolution of fossil plants. *Elsevier Press*, 1-1230.

Tekleva M, Markevich V, Bugdaeva E, et al. 2015. *Pseudointegricorpus clarireticulatum* (Samoilovitch) Takahashi: Morphology and ultrastructure. *Historical Biology*, 27(3–4): 355-365.

Tekleva M, Polevova S, Bugdaeva E, et al. 2019. Further interpretation of *Wodehouseia spinata* Stanley from the Late Maastrichtian of the Far East (China). *Paleontological Journal*, 53(2): 203–213.

Tekleva M, Polevova S, Bugdaeva E, et al. 2020. Three *Aquilapollenites* species from the late Maastrichtian of China: New data and comparisons. *Rev. Paleobot. Palynol.*, 282: 104288. hppts://doi.org/10.1016/j.revpalbo.2020.104288.

Vachrameev V A. 1988. *Jurassic and Cretaceous floras and climate of the Earth. Proceedings of the Geological Institute of the USSR, 430*. Moscow: Nauka, 1-209. (in Russian)

Wan X, Zhao J, Scott R W, et al. 2013. Late Cretaceous stratigraphy, Songliao Basin, NE China: SK1 cores. *Palaeogeography, Palaeoclimatology, Palaeoecology*, 385: 31-43.

Wang H S. 2002. Diversity of angiosperm leaf megafossils from the Dakota Formation (Cenomanian, Cretaceous), north western interior, USA. Thesis (Ph. D.). University of Florida, 1-204.

Wang H S, Dilcher D L. 2006. Angiosperm leaf megafossils from the Dakota Formation: Braun's Ranch Locality, Cloud County, Kansas, USA. *Palaeontographica* B, 273: 101-137.

Wang Y D, Huang C M, Sun B N, et al., 2014. Paleo-CO_2 variation trends and the Cretaceous greenhouse climate. *Earth-Science Reviews*, 129: 136-147.

Wolfe J A. 1995. Paleoclimatic estimates from Tertiary leaf assemblages. *Ann. Rev. Earth Planet Sci.*, 23: 119-142.

Wolfe J A, Spicer R A. 1999. Fossil leaf character states: Multivariate analysis. In: Jones T P, Rowe N P (eds.) *Fossil Plants and Spores: Modern Techniques*. Geological Society, London, 233-239.

Wu W H, Godefroit P, Han J X. 2010. A hadrosaurine dentary from Upper Cretaceous of Jiayin, Heilong-jiang. *Global Geology*, 29(1): 1-5. (in Chinese)

Wu W H, Zhou C F, Wing O, et al. 2013. A new gigantic sauropod dinosaur from the Middle Jurassic of Shanshan, Xinjiang. *Global Geology*, 25(3): 1-8. (in Chinese)

Xi D P, Wan X Q, Li G B, et al. 2018. Cretaceous integrative stratigraphy and timescale of China. *Science China* (*Earth Sciences*). https://doi.org./10.1007/s1430-017-9262-y.

Xi D P, Wan X Q, Li G B, et al. 2019. Cretaceous integrative stratigraphy and timescale of China. *Science China* (*Earth Sciences*). 49(1): 257-288. (in Chinese with English abstract)

Zhang W, Zheng S L, Zhang Z C. 1980. Paleontological atlas of Northeast China (2). Beijing: Geological

Publishing House. 221-307. (in Chinese)

Zhang Z C. 1981. Several Cretaceous angiospers from Mudanjiang basin, Heilongjiang. *Bull. Shenyang Inst. Geol. Min.*, 2(1):1-9. (in Chinese with English abstract)

Zhang Z C. 1984. The Upper Cretaceous fossil plant from Jiayin region, northern Heilongjiang. *Proceeding of Stratigraphy and Paleontology*, 11: 111-132. (in Chinese with English abstract)

Zhang Z C. 1985. Main stages of Cretaceous angiosperm succession in north part of Northeast China. *Act. Palaeont. Sin.*, 24(4): 453-460. (in Chinese with English abstract)

Zheng S L, Zhang Y. 1994. Cretaceous plants from Songliao basin, Northeast China. *Act. Palaeont. Sin.*, 33 (6): 756-764. (in Chinese with English abstract)

Zhou Z Y, Li P J. 1980. A paleobotanical approach to the classification, correlation and geological ages of the non-marine deposits of China. In: Scientific papers on geology for international exchange, prepared for the 26th IGC. 4. Beijing: Geological Publishing House, 82-91.

孙　革　古植物学家，沈阳师范大学/吉林大学教授、博士；国家古生物化石专家委员会顾问、中国古生物学会副监事长、中国古生物学会古植物学分会名誉理事长兼科普工作委员会名誉主任，第六届国际古植物学会（IOP）副主席。1968年毕业于长春地质学院（今吉林大学），1985年于中国科学院南京地质古生物研究所获博士学位，1988~1989年英国大英博物馆（自然史部）博士后。专长于中生代植物、早期被子植物及事件地层学研究，曾率课题组首次发现迄今世界最早的被子植物辽宁古果及中华古果，首次提出"被子植物起源的东亚中心"假说，上述成果曾以封面文章在美国《科学》（*Science*）杂志上发表。已发表论文200余篇、专著8部，包括《辽西早期被子植物及伴生植物群》（2001）及《黑龙江嘉荫晚白垩世—古新世生物群、K-Pg界线及恐龙灭绝》（2014）等。出版译著1部、参与撰写专著4部。曾获教育部自然科学一等奖（2004）、李四光地质科学奖（2005）及辽宁省科学技术一等奖（2014）等。2014年获全国优秀科技工作者称号。

梁　飞　沈阳师范大学副教授,博士,硕士生导师,2015年于吉林大学获古生物学与地层学博士学位,同年起在沈阳师范大学古生物学院任教。主持国家自然科学基金"黑龙江嘉荫晚白垩世永安村组植物群及其地层对比"项目;率课题组首次发现晚白垩世水生被子植物"嘉荫莲"化石;2019年入选辽宁省"百千万人才工程"计划。

杨　涛　沈阳师范大学副教授,博士,硕士生导师,2004年毕业于吉林大学地球科学学院,2010年于吉林大学获古生物学与地层学博士学位,同年起在沈阳师范大学古生物学院任教。专长于晚古生代安加拉植物群及中生代植物群与地层研究,2006年起参加黑龙江嘉荫晚白垩世植物群及地层研究工作。

张淑芹　吉林大学研究员,1984年毕业于长春地质学院(现吉林大学)地层古生物专业,2006年由中科院东北地理与农业生态所调入吉林大学古生物研究中心工作,专长于新生代孢粉与地层研究;近年来先后在内蒙古林西二叠系、吉林东部及松辽盆地白垩系、黑龙江嘉荫K-Pg界线地层,以及长白山新生代地层和古植被、古气候等研究中,取得多项重要研究成果。2015年起参加黑龙江嘉荫晚白垩世植物群研究工作。